SCIENTIFIC PAPERS MADE EASY

Scientific Papers Made Easy

How to Write With Clarity and Impact in the Life Sciences

Stuart West

Professor of Evolutionary Biology, Department of Biology, University of Oxford

Lindsay Turnbull

Professor of Plant Ecology, Department of Biology, University of Oxford

OXFORD
UNIVERSITY PRESS

OXFORD
UNIVERSITY PRESS

Great Clarendon Street, Oxford, OX2 6DP,
United Kingdom

Oxford University Press is a department of the University of Oxford.
It furthers the University's objective of excellence in research, scholarship,
and education by publishing worldwide. Oxford is a registered trade mark of
Oxford University Press in the UK and in certain other countries

Published in the United States of America by Oxford University Press
198 Madison Avenue, New York, NY 10016, United States of America

British Library Cataloguing in Publication Data

Data available

Library of Congress Control Number: 2022945704

ISBN 978–0–19–286278–5 (hbk)
ISBN 978–0–19–286279–2 (pbk)

DOI: 10.1093/oso/9780192862785.001.0001

Acknowledgements

This book has been improved enormously by feedback from all those who have read chapters, or who have taken one of our courses on scientific writing. We especially thank: Louis Bell-Roberts, Saran Davies, Anna Dewar, Cecilia Karlsson, Asher Leeks, Ming Liu, Alper Mutlu, Mati Patel, Tom Scott and Josh Thomas for detailed feedback during a graduate course; Ashleigh Griffin for suggesting exercises; Max Burton for commenting on the whole book and providing a useful acronym. Pandora Dewan drew the amazing cartoons. Anna Dewar heroically prepared all the figures for chapter five.

Contents

1

Writing as an Essential Research Skill

The impact of scientific research depends critically on how well it is communicated to others. In most sciences, this means writing a scientific paper that other scientists will read. Perhaps you have done amazing experiments and think that this will guarantee success. However, without good writing, you may struggle to get your paper published and your brilliant experiments will not have the impact that they deserve. Good writing can't save bad science, but bad writing can sink good science (Cartoon 1.1).

Cartoon 1.1 *The weakest link. Science can be imagined as a chain with three links. These links represent: (1) identifying a problem that needs solving; (2) doing the research, such as an experiment, to solve that problem; and (3) communicating the results, primarily through writing a peer-reviewed publication. Ultimately, a piece of science is only as good as its weakest link, so however strong the first two links, if the writing is weak, the chain will break.*

Most scientists accept that learning new skills is an essential part of carrying out good research. This might involve acquiring new laboratory techniques or slaving over a new computer program. But while most scientists accept the time required to learn skills, it is often assumed that writing is something that can just be done, because it was learnt at school or university. Indeed, the writing of a scientific paper is often seen as an undemanding task that can be done quickly at the end, once the real job of doing the research is finished.

In our opinion, this view of scientific writing could not be further from the truth. The writing of a scientific paper is hard work. Furthermore, writing is a skill that needs to be learnt and practised. Indeed, good writing can require a much longer learning period

Scientific Papers Made Easy. Stuart West and Lindsay Turnbull, Oxford University Press. © Stuart West and Lindsay Turnbull (2023).
DOI: 10.1093/oso/9780192862785.003.0001

than many familiar research techniques. Although good writing is hard, we believe that a few simple tips can easily remedy many of the most familiar pitfalls, and the good news is that scientists keep improving their writing throughout their careers.

In this book we aim to help you produce better scientific papers. We accept that writing is a high-level skill that can't be mastered by just reading a guidebook, and that it's also personal—your style is particular to you. However, we believe that a few simple pointers would improve the quality of much writing. We see the same mistakes time and time again—mistakes that could easily be fixed. By helping avoid these errors, we hope to make the experience of writing more satisfying, less daunting, and, dare we say it, even fun.

Our fundamental principle is simple: *The reader must come first*. Too often scientists write for themselves or to sound scientific, when in fact their job is to help the reader understand the content of their paper by making things as clear and straightforward as possible (Box 1.1). The story of your paper must be easy to digest and written in relatively plain English, rather than in highly technical gobbledygook. Strangely, this does not always come naturally, and requires a change of thinking on the part of the writer. For example, it requires the writer to ask questions like: is there any possible way that this sentence could be misunderstood? We believe that, once mastered, the rewards of taking our approach are enormous, as your science will have so much more impact.

We have written this book in the form of a toolkit for producing the different sections of a standard scientific paper: Abstract, Introduction, Methods, Results, and Discussion. Each chapter stands alone, so that you can read them in any order, or dip in and out, as required (Box 1.2). There are also boxes and summaries with helpful advice for those who like to flick through books and don't want to read this one from cover to cover. At the end of each chapter there is a summary table, so if you want to return to a particular chapter, you don't have to read the whole thing again. We hope that by breaking a paper down into manageable sections, you will find it easier to write. We have also provided supplementary online material to complement the chapters, and to facilitate running a course (class) based on this book (Box 1.3).

Getting started on a scientific paper is often the hardest part. People have different views on where to start, but we present the sections of a scientific paper in the order in which we think it is easiest to write them. Methods are a good place to start because you simply describe what you did. The Results section follows naturally from the Methods, so we present this chapter next. The Introduction and Discussion are harder to write, as they require context, links to the literature, and generally more thoughtful analysis and interpretation. Counter-intuitively, although the Abstract is the first section of a finished scientific paper, we think it is best written last, once you are happy with everything else and are clear about the story of your paper. Finally, after covering the different sections of a paper, we explain how to write a cover letter, and how to edit a paper, once you have a full draft.

We hope that you are now itching to get started, but before we get on to the details of specific sections, we provide some tips that help all sections. Take the core skills quiz before reading Chapter 2 (Box 1.4).

Box 1.1 The Reader

Writers often imagine that readers will diligently work through their paper, section by section, until they have read and understood every word. But, while this was often the case at school and university, there's no guarantee that potential readers of scientific papers will apply the same level of diligence. There is a rapidly expanding literature within every scientific discipline and finding time to read the latest research is increasingly challenging for scientists at all levels.

A far better image of your potential reader is someone time-limited, stressed, and easily bored (Cartoon 1.2). They have a million other things to do and will take any excuse to give up on reading your paper. They might be a PhD student trying to get to grips with their subject, or a professor who doesn't really have time to read papers anymore. Unfortunately, unless they are a reviewer, they don't *have* to read your paper, so it's your job to make them want to.

Cartoon 1.2 *The reader. Picture your potential reader. You are competing for their attention with other scientific papers, other work tasks, email, Twitter, the entire content of the internet, and any other hobbies or pastimes that they enjoy. Getting someone to read and understand your paper shouldn't be taken for granted; it is an incredible achievement.*

To convince potential readers to keep reading your paper, you must write it in a way that they can easily understand. They might not be quite as interested in your subject area as you are. They might be tired, hungover, or just in a bad mood. Indeed, even referees who have agreed to give your paper a fair chance will not enjoy struggling through a poorly written paper to find its inner beauty. If referees find a paper confusing or hard, then they might just reject it. The Journal of *Environmental Microbiology* published a few quotes from real reviewers to highlight this problem:

'*The biggest problem with this manuscript, which has nearly sucked the will to live out of me, is the terrible writing style.*'

'*The writing and data presentation are so bad that I had to leave work and go home early and then spend time to wonder what life is about.*'

While these comments are somewhat extreme and were chosen to entertain readers of the journal, they emphasize how much good writing can matter.

When their papers are rejected, authors often curse referees for being idiots who almost wilfully chose to misunderstand their paper. But that irritating reviewer wasn't a randomly selected person—they were chosen by the editor as someone who could reasonably assess the author's work. The referee felt that they knew enough to accept the refereeing assignment, despite having a million other things to do. Consequently, they are a good approximation to the kind of reader that the author wants to attract.

So, remember, it's your job to make your paper easy for potential readers to understand. It's not the reader's job to struggle their way through something that they can't follow.

Box 1.2 Is This Book for You?

We have targeted this book at researchers across the biological, life, and human sciences. However, we believe that the core points apply more widely to any natural science, and to other related disciplines such as applied mathematics or computer science. We illustrate our points with simple biological and human examples that do not assume prior knowledge and could be understood by anyone with a scientific background. Producing readable papers is especially important for interdisciplinary science. The top scientific journals expect their papers to be understood by a broad range of scientists, including non-specialists. Our book is aimed at both native and non-native English speakers, with applied, actionable tips, and numerous illustrative examples of good and bad writing.

Box 1.3 Supplementary Online Material

We have also provided additional material, to help readers of this book, and to facilitate using this book to teach a course on writing. We have developed some short quizzes, which we recommend you take before reading each chapter, and some simple exercises to illustrate or reinforce principles. We have developed a scientific writing course based on this book. This course involves eight classes, where the students work in groups to complete exercises before each class. We have taught this course at Oxford, to both Masters and PhD students. All the supplementary material can be reached via signposted QR codes, or from the companion website (https://global.oup.com/us/companion.websites/978019286278/sm/).

Box 1.4 Take the Core Skills Quiz before reading Chapter 2

2

Core Skills

There are very few rules about writing that can't be broken. Skilled writers often play with rules or conventions. But this book isn't aimed at skilled writers—it is aimed at those who are just starting, or are trying to make their writing better.

In this chapter we provide a few core skills to help new and improving writers. Most points are better illustrated in the context of specific paper sections, and we don't want to front-load this book with too many guidelines that need to be remembered. Consequently, in this chapter, we focus on a small number of points that can be applied in every section of a paper.

We provide:

- four general principles;
- our top 10 specific tips.

Remember—these are not hard and fast rules, just tips to make scientific writing easier and clearer. We return to all of them in later chapters, when discussing how to write specific sections of a paper. The uniting theme is to produce simple, uncluttered writing with a clear narrative flow.

Four General Principles

Keep it simple

When students first start writing papers, they often seem to take on a style that they think makes them sound like a scientist. This can include using jargon, technical abbreviations, and phrases that they would never use in everyday speech. But this kind of writing makes things harder, not easier, for potential readers, and may discourage them from carrying on with your paper.

It is your job to attract a broad range of readers to your paper and to make it as easy as possible for them to understand what you found out. Simple, clear writing is fundamental to this task. Instead of trying to sound scientific or clever, you should be clear and concise. This is not easy—it can sometimes be harder to explain scientific

Scientific Papers Made Easy. Stuart West and Lindsay Turnbull, Oxford University Press. © Stuart West and Lindsay Turnbull (2023).
DOI: 10.1093/oso/9780192862785.003.0002

concepts using plain English than by using technical language. But, if you persevere, it will start to come more naturally. In the later chapters we provide numerous examples of overcomplicated writing, and how to make it simpler.

Assume nothing

Step back and put yourself in the position of the reader. To understand your paper, the reader needs an explanation of all the necessary steps. But 'necessary' only makes sense in the context of what they already know. And it's easy to assume that your reader knows much more about the topic than they actually do.

Even an expert in your subject lacks specific knowledge about what you have done and why you have done it. In contrast, you have been thinking about your work for months—maybe even years—on an almost daily basis. You will be incredibly knowledgeable about every detail of your work: from the other studies that inspired you, to what you actually did and found out—and its implications. You will have forgotten all the hours that you invested in understanding these different aspects of your work. Indeed, you are probably the worst person to write up your study, because you are as far away as it's possible to be from someone who is new to the topic.

This problem is so common that it has even been given a name in the psychology and economics literature: the 'curse of knowledge'.[1] The curse means that, when communicating with others, people tend to unwittingly assume that others have the necessary background to understand what they are saying. It's too easy to miss out crucial points or steps because you have forgotten how important they are for understanding your work.

So, how can we tackle the curse of knowledge? When writing a paper, it's best to assume that your reader is scientifically literate, but has very little expert knowledge. Your paper is more likely to fail because you assumed too much, than because you dumbed it down too much. This is especially true for interdisciplinary science, where different readers can have completely different backgrounds.

Remember—you need to go through all of the crucial points, step by step, taking your reader with you. In the following chapters we will provide a toolkit to help you do this, by providing structures for the different paper sections that force you to cover the key points needed for the reader.

Keep it to essentials

Let's assume that you know a lot, and have opinions about many things. But which facts and opinions do you really need to include in your paper? Is a digression on topic X or Y necessary or relevant? Authors are often tempted to add more to their papers to show how much they know or to give their opinion about something. But this isn't an opportunity to tell the world what you know or think about a broad range of topics.

Instead, we suggest that you strip your scientific writing down to the bare essentials (minimalism). This doesn't mean making the paper so short that it doesn't do its job

---

[1] Camerer *et al.* 1987. The curse of knowledge in economic settings: an experimental analysis. *Journal of Political Economy*. 97: 1232–1254

properly. It's important to include all the information that's *necessary* to understand the paper—but no more.

Editing a paper so that it includes 'just enough' is actually a lot harder than writing a longer paper. But it's worth the effort, because a shorter paper helps the reader. If you focus on the main message, and remove all distractions, then the reader will come away with the message that you want them to have. Unfortunately, most readers will only ever absorb a very small fraction of what you have written. So try to hammer home the messages that you most want your reader to take away. Brevity also minimizes the likelihood that you lose the attention of your time-stressed reader. People are more likely to start reading—and to finish—shorter papers.

Tell your story

Good scientific writing tells a story. It tells the reader why the topic you have chosen is important, what you found out, and why that matters. For the story to flow smoothly, the different parts need to link clearly to each other. In creative writing this is called 'narrative flow'.

If your paper has narrative flow, the reader will be gently led from one section to the next, excited to find out what happens next. The main points will be clear and they will grasp them. If your writing doesn't have narrative flow, the reader will get confused—and very possibly frustrated.

This advice might seem rather vague at the moment, but in the later chapters we will provide a structure for each chapter, which will lead to narrative flow.

Top 10 Tips

Write as you speak

Scientists often tend to write phrases that they would never say, possibly because they think they need to 'sound scientific'. Instead, you should explain your science in the clearest possible way, using simple English whenever possible. If you never say the word 'whilst' when speaking to friends, then don't use that word in your paper. *Keep it simple.* Compare:

1. *We assigned birds to two experimental groups, where we manipulated the nutritional resources to different levels.*
2. *We divided the birds into two groups, which received small or large amounts of food.*

1. *These results suggest that worms act as rational agents, able to increase their Darwinian fitness by adjusting their basic reproductive rate conditionally in response to environmental conditions.*
2. *The results suggest that when there is more food available, worms lay more eggs.*

Cartoon 2.1 *The robot test. A good way to 'test' your writing is to read it out loud. This is an easy way to reveal common errors, such as too much jargon, or sentences that are too short or too long. Often, bad writing will make you sound like a robot.*

In both examples, sentence 2 is written in simpler English, and is shorter and easier to understand. A good test of whether you have got it right is to read your sentence out loud (Cartoon 2.1). In both of the examples above, it's hard to not sound like a robot when reading out sentence 1, whereas sentence 2 will sound more natural.

Write short sentences, which only make one point

Short sentences with clear and simple constructions are easier to read and grasp. Readers normally hold a whole sentence in their head, as they try to make sense of it. So, if you put too much into one sentence, they will find it hard to process and may have to re-read it multiple times. Consequently, put one idea/point per sentence, and consider breaking a long sentence into two or more shorter ones. Compare:

1. *After harvesting all plants and drying them in an oven at 80°C for 24 hours we weighed them on a microbalance to the nearest 0.1 g and then carried out CHN analysis to determine the proportions of major elements after which we ran small samples through HPLC to obtain a full chemical profile.*

2. *After harvesting all plants and drying them in an oven at 80°C for 24 hours we weighed them on a microbalance to the nearest 0.1 g. To determine the proportions of major elements we carried out CHN analysis. To obtain a full chemical profile we ran small samples through HPLC.*

1. *As a control, we made all individuals play a single round of the same game but with computerized groupmates, rather than with real humans, in order to test whether their behaviour depends upon concerns for the welfare of others.*

2. *As a control, we made all individuals play a single round of the same game but with computerized groupmates, rather than with real humans. This control tests whether behaviour depends upon concerns for the welfare of others.*

As a rule of thumb, if a sentence is >2 lines long in something like Microsoft Word, then alarm bells should go off. Look at that sentence and think about whether it needs to be split into multiple sentences. Or read it out loud—it will quickly become obvious if you have put too much in. Of course, at the same time, don't make too many of your sentences too short—we will return to this point in Chapter 3.

Avoid jargon

A crucial part of writing in simple understandable English is to avoid jargon. Other people might use jargon differently, or not understand it all. People sometimes use jargon to sound clever or scientific—or maybe they even think they have to. But jargon puts readers off. This problem has even been quantified—papers with more jargon are less likely to be cited.[2]

It can be harder and more work to write a paper in simple English, without jargon, but your paper will have a wider audience. Compare:

1. *In microbial systems, intraspecific competition is induced by stressful conditions such as resource or space limitation.*
2. *Bacteria compete for space and resources.*

1. *The evolutionary legacy hypothesis proposes that an evolved reciprocity-based psychology affects human behaviour in anonymous one-shot interactions when reciprocity is not explicitly possible.*
2. *It has been hypothesized that humans cooperate more if they feel they are being watched.*

1. *We listed all the information regarding the full genome analysis of the strains used in this competition assay in Table 1.*
2. *We listed all mutations in Table 1.*

In sentence 2, the jargon from sentence 1 has been replaced with simple English, making it much easier to understand. Be warned—it's easy to get sucked into writing jargon—each sentence 1 is taken from a real paper.

Sometimes avoiding jargon is relative. For example, in fields such as immunology, there is a lot of jargon that cannot be avoided. When writing an immunology paper, avoiding jargon means avoiding terms that aren't broadly understood by immunologists. Similarly, most fields of research will involve some standard terms that are used by

---

[2] Martinez & Mammola. 2021. Specialized terminology reduces the number of citations of scientific papers. *Proceedings of the Royal Society London Series B* 288:20202581

everyone. The general tip is to avoid terms that will only be known by a narrow set of specialists. In this book we have deliberately used simple illustrative examples that can be understood by all readers.

At other times you might want to use jargon, to make links to other literature clear. In this case, you can still write in simple English, but then add the jargon, either in brackets or by explaining it. For example:

1. *Humans often face opportunities to improve the welfare of their group, but at a cost to the individual ('social dilemmas').*

2. *Humans often face social dilemmas, where they can improve the welfare of their group, but at an individual cost.*

Write cause before effect

Many sentences in scientific papers contain a cause and an effect. You might describe what you did (cause) and the consequence (effect), or variation in something (cause) and what this led to (effect). You can help the reader by putting the cause first, followed by the effect. Compare:

1. *More matings with females are obtained by male peacocks with larger tails.*
2. *Male peacocks with larger tails mated with more females.*

1. *Females laid more eggs when they were larger.*
2. *Larger females laid more eggs.*

1. *Supernatants contained no live or dead cells, only exoproducts produced by the cells in the original culture, as they were produced using a 0.22 μm filter.*
2. *The supernatant was produced with a 0.22 μm filter, and so contained no live or dead cells, only exoproducts produced by the cells in the original culture.*

In all cases, sentence 2 presents the cause first, followed by the effect. This helps because the cause explains the effect. If you do it the other way around—effect and then cause—the reader might have to go back and reassess the effect after reading the cause, perhaps by rereading the sentence. Writing cause before effect makes the sentence easier to process.

Make one major point per paragraph

Each paragraph should make one major point. The purpose of paragraphs is to break the text up into manageable units, so that the reader can grasp the main points. If a paragraph contains just one major point, then it will stand out clearly. If it contains more than one, then the main message is lost, and the reader can become confused. A typical size for a paragraph would be three to eight sentences. We will provide examples of good and bad paragraphs in later chapters, when discussing specific sections.

Use the first sentence of a paragraph to summarize that paragraph

It is often useful to make the first sentence of a paragraph a summary of the major point of that paragraph. The rest of that paragraph then provides the details, which could range from a detailed explanation to statistical analyses. The first sentence gives the reader the key point up front, and it tells them what that paragraph is going to be about. They then know where they are going and will be at ease.

If done well, the collection of first sentences will provide a useful summary of your paper. To test yourself, once you have written a section, or a whole paper, read just the first sentences of each paragraph. Ask yourself whether a reader that only read those first sentences would obtain a reasonable understanding (Box 2.1). The answer should be *yes*.

Readers will often lose concentration in a paragraph or just skim through a paper. If you summarize each paragraph with the first sentence, then they will still get the main points, even if that is all they read. Make it as easy as you can for your readers.

Box 2.1 Complete the First Sentence Exercise

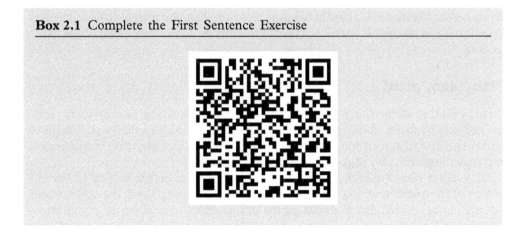

Be precise and consistent

It's of paramount importance to be precise, especially with terms and definitions. This sometimes means that you have to sacrifice variety of language for precision. For example, if you have done an ecology field experiment, and you begin by stating that you have 'plots' of 2 m by 2 m, then for the rest of the paper you must always refer to them as plots. Don't sometimes call them 'patches'. It's surprisingly easy to use different words to refer to the same thing in different parts of your paper, so you need to actively avoid this.

You should also be consistent with terminology across the existing literature. If other papers use a term in a specific way, and you use it differently, it will spread confusion and hinder progress.[3] Science is precise. Scientific progress depends upon precise, reliable communication between scientists.

Follow the section headings

The heading of each section of a paper provides a major clue as to what you should include. The Introduction is for introducing why your paper is needed, the Methods is for methods, and so on. Having separate sections helps the reader, because they know what to expect in each section. In addition, the different sections provide information in a logical order: why you carried out your study (Introduction), what you did (Methods), what you found out (Results), and what those findings mean (Discussion). Each of these sections follows on nicely from the previous one, and leads neatly into the next.

If you put the 'wrong' bits in a section then you are either being repetitive or you didn't do a good enough job in the appropriate section. For example, your Discussion section shouldn't start with a long introduction as to why your work was needed—that belongs in the 'Introduction'. Putting things in the wrong section can confuse or frustrate your reader.

Following the section headings doesn't mean that there can be no repetition of elements between sections. Linking between sections can help the reader. In later chapters we will suggest when it is useful to repeat elements or summarize points from other sections.

Plan, plan, plan!

It can be hard to start writing. A blank page can be very daunting. Some writers tackle this problem by focusing on 'getting something down'. The idea here is that it helps to get started, and that once you have something down, you can play with it and hone it into something great. We suggest the opposite approach.

We suggest planning each section as much as possible, before writing it. For the section of the paper that you want to write, begin by writing down the bullet points that need to be covered. For example, for the Introduction, write down the key information about why your paper is needed, or for the Methods, write down the essential steps in your protocol.

It is often better to plan with pen and paper. You can then play with the different bullet points, ordering and linking them with a flow diagram (mind map) to produce

---

[3] West *et al.* 2007. Social semantics: altruism, cooperation, mutualism, strong reciprocity and group selection. *Journal of Evolutionary Biology* 20: 415–432.

Cartoon 2.2 ***Plan, plan, plan!*** *It is often useful to plan with a pen and paper.*

the structure of each section (Cartoon 2.2). Add extra bullet points as needed. Each bullet point will usually correspond to one paragraph and provide the major point of that paragraph. Given that the first sentence of each paragraph should summarize that paragraph (page 13), your bullet points will be very close to the first sentence of each paragraph.

Once you have the first sentence of each paragraph, you just need to add the rest of each paragraph by elaborating on that first sentence. And once that is done, you will have finished writing that section of your paper! In the later chapters of this book, we will provide more specific toolkits for planning the structure of each section of a paper.

Our emphasis here is to separate the planning from the actual writing. We suggest doing this for three reasons. First, if you jump straight in, you are likely to produce a disorganized stream of consciousness that is so useless it is probably better to throw it away and start again anyway. Second, planning in advance allows for much clearer thought: you just need to plan and organize the big picture, without getting lost in the details. Third, once you have actual text down, it can be hard to change it—you have made the effort to type it in, so you don't want to just throw it away. This is also so well known that psychologists have a phrase for it: the 'sunk cost fallacy'.

The method of building up from a flow diagram might feel slow, but it can be easier and faster in the long run. If you start by building up a plan, and by gradually fleshing it out, you will end up with something much closer to a final version. In the end, this means fewer iterations and corrections. You will still have to tweak your paper and improve it, but you are unlikely to have to throw it out and start again.

Good planning can also help with the curse of knowledge. On page 8, we emphasized the importance of stepping back and putting yourself in the reader's place. Planning involves stepping back and thinking about the big picture. At this stage, it's much easier to think about what you need to explain to the reader. Consider all the logical steps that are obvious to you, but might not be obvious to someone who hasn't spent the last

Cartoon 2.3 **The golden rule.** *You can ignore tips—do whatever helps your reader.*

few years buried in the topic. *Good planning will help you write faster, and produce better papers.*

You can ignore the above tips

Following the tips in this book will generally lead to clearer writing. But we cannot emphasize enough that they are only tips, not hard and fast rules. Sometimes it can be better not to follow them—the golden rule is that the reader must come first (Cartoon 2.3). *Do whatever helps your reader!*

To give an example, it can sometimes be useful to deviate from Top Tip 4, and put the effect before the cause. Compare:

1. *Bacterial pathogens produce toxins to eliminate competitors, allowing them to grow better.*
2. *The growth of bacterial pathogens depends upon their ability to eliminate competitors with toxins.*

Sentence 2 puts the effect before the cause, but this order might work best if we wanted to put the focus on pathogen growth. This can depend upon the purpose of the sentence within the context of the overall section. However, you don't have to agree with us—the point is that there is no absolutely wrong and right way to do this. It's simply important to think about different possibilities, and to do what you think will make things easiest for your reader. We have tried to follow our own recomendations in this book, but we are fallible, and so will have occasionally failed.

Summary

Great guide to wonderful writing
General principles
1. Keep it simple
2. Assume nothing
3. Keep it to essentials
4. Tell your story
Top 10 tips
1. Write as you speak
2. Write short sentences, which only make one point
3. Avoid jargon
4. Write cause before effect
5. Make one major point per paragraph
6. Use the first sentence of a paragraph to summarize that paragraph
7. Be precise and consistent
8. Follow the section headings
9. Plan, plan, plan!
10. You can ignore the above tips

Before reading Chapter 3, summarize the main points of this chapter (Box 2.2) and take the Methods quiz (Box 2.3).

Box 2.2 Mind Maps

After reading this chapter, make a mind map to summarize the points. Try to make it in a way that would provide a useful summary, to save you having to read this chapter again. Repeat this exercise for each chapter, as you work through the book.

Box 2.3 Take the Methods Quiz before reading Chapter 3

3

Methods

The Methods section is usually the best place to start writing. After all, you simply tell everyone what you did and you should be clear on that (Cartoon 3.1). But an endless list of detailed protocols is difficult to digest. So, how do you persuade your reader to keep going? And which details matter?

Cartoon 3.1 *Methods. The Methods section is for telling everyone what you did.*

The key to a digestible Methods section is to supply information at a carefully controlled rate (Cartoon 3.2). There are two parts to this process.

- First, pull out the essential information and place it prominently up front. This provides an overview and creates a structure in which to place the forthcoming details.
- Second, when you are providing the details, keep reminding the reader exactly *why* you are telling them something. Information in context is much easier to digest. If you leave your reader drowning in what appear to be unnecessary details, they might just give up.

Our first aim in this chapter is to show you how to structure a Methods section, dividing it into an overview and the details. We then examine how to fill in the details, by making use of the tips that we gave in Chapter 2. These are the core skills needed to produce any Methods section.

Scientific Papers Made Easy. Stuart West and Lindsay Turnbull, Oxford University Press. © Stuart West and Lindsay Turnbull (2023).
DOI: 10.1093/oso/9780192862785.003.0003

Cartoon 3.2 *Digestible methods. Supply information at a controlled rate, by dividing it between an overview and the details.*

Finally, we provide guidance on how to deal with different types of paper. Methods sections are highly variable, depending on the type of work carried out. You might have performed a simple and elegant experiment, collected a huge observational data set that needs a complicated analysis, or developed a mathematical model. Here, we give advice for writing all of these different types of Methods sections. Whenever possible, we have used boxes, so that you can more easily skip the bits that are not relevant to the paper you are writing.

Structuring Methods

Most writing advice will tell you that a good Methods section should allow someone else to come along and perfectly replicate what you did. But, if you follow this advice too slavishly, you will end up with a Methods section that is far too long and unwieldy for the average reader. And how many of your readers will really want to replicate your work?

The typical reader only really needs enough information to do three things:

(1) they need to broadly understand what you did . . . *so:*

(2) they can make sense of your results and judge their quality . . . *so:*

(3) they can critically evaluate the interpretation that you are choosing to make.

But a few readers, including reviewers, need to know more. So how can you cater to these two very different kinds of readers?

As with all sections of a paper, there's no single solution, but here's a good way to start. Separate your Methods into an overview and the details:

> The Overview: The overview is one paragraph at the start of your Methods that summarizes its key points and gives an impression of the scope, scale, and quality of your study. It orients the reader, and allows them to place the details into context.
>   The Details: The bulk of your Methods section is a series of paragraphs with clear subheadings, which spell out exactly what you did.

The Overview

The overview should supply the reader with the key points of your study. It allows them to orient themselves and gives a sense of the scope, scale, and quality of your work. It must include a rationale—what was the logic behind the study? Of course, this information is laid out in detail in the Introduction, but we can't assume that the reader will read all of your paper! Maybe they skipped from the Abstract to the Methods, or maybe they are just dipping back in again after initially reading the whole paper a few months ago.

A good overview should allow a casual reader to then jump straight to the Results and make sense of the figures they will see there. Obviously, reviewers can't do this, but, once a paper is published, many casual readers are happy to trust that the detailed methodology is probably sound. A good working hypothesis is that after reading your overview, most readers skim, or even ignore, the rest of the Methods section. Make your overview good enough that they can get away with that!

Here are some suggestions for the kind of information that you might like to include in your overview:

- type of study—whether experiment, observation, modelling, or a combination of all three
- place it was carried out
- duration
- the study system—this might be the names of any study organisms involved, the genes examined, or how you got human volunteers
- the layout of any experiment
- the type of data collected or analysed
- modelling approach taken—analytical, numerical, or simulation
- the key assumptions of a theoretical model

Once you have decided which key points to include in your overview, you need to present these in a clear and comprehensible way, in plain English and without jargon. People who aren't deeply familiar with your field might nevertheless want to look at your paper. So, why not make it easy for them to do so?

To see how to do this, let's start with a very simple example of an ecological field experiment.

> *Overview:* To test how grass communities respond to increased soil nitrogen, we conducted a field experiment in which we experimentally added nitrogen to plots. To cover the range of nitrogen inputs commonly seen across lowland England, we used three levels of nitrogen addition plus a control. We marked out 20 plots across the site and randomly assigned the four treatments to plots, giving five replicates of each treatment. We applied nitrogen three times during the growing season for three consecutive years. We measured the response of the grass communities by sampling total biomass and species composition in July each year.

Why do we think that this overview works? First, the reader knows *why* this study was conducted—to test how grass communities respond to increased nitrogen. Second, they know *what* the authors did: they added nitrogen to a natural grass community. Third, the authors placed their work in a realistic context—they conducted a field experiment and used sensible levels of nitrogen inputs. Fourth, the design of their experiment is clear, in that we know: (1) the number of treatments and the sample size, (2) that treatments were assigned randomly to plots, and (3) the experiment included a control. Fifth, the authors emphasize that this was not a single-season study, but continued for three years. This is likely to be an important indicator of quality, because effects might take time to develop. Finally, they have identified the response variables—biomass and species composition—so the reader knows what they are going to find on the *y*-axis of upcoming figures.

The language of this overview is also clear and non-technical. It does not include jargon or acronyms, which might put off a non-specialist. It does not include details about exact amounts of nitrogen or plot sizes. This overview allows a more specialist reader to start formulating their own questions about forthcoming details that they are expecting to see—for example, what exactly are the three rates of nitrogen addition? If they encounter these details in later paragraphs, then this gives them a sense of satisfaction.

Perhaps most importantly, this short paragraph enables a reader to visualize the study: 20 plots laid out in a grassland, receiving different amounts of nitrogen for three years. It's surprising what a difference this can make. We read a lot of papers where, halfway through the Methods, we still find ourselves asking: *what* did they do? That is frustrating. Contrast with the happy reader of your paper who will be thinking 'that was an awesome overview, now I can just skip straight to the results'.

Our suggested overview contrasts with how the Methods section of many papers start. For example, ecology papers often begin with a detailed description of the study site—information that is hard to process out of context. But, if an overview is given first, the details, such as a description of the study site, will make more sense to the reader.

Each study will be different, so you'll have to think carefully about what to include in your overview and what can be left for the details. We provide a couple more overview examples in Box 3.1. Don't forget, it's not about hiding details, or sweeping them under the carpet (Cartoon 3.3). The details *do* matter. It's just that they are much easier to digest once the overview is clear.

Cartoon 3.3 *Don't hide details!*

Box 3.1 Different Overviews for Different Subjects

We think that an overview can be useful whatever your subject area. It's true that in some fields, such as molecular biology, there is a lot more technical language and the details are particularly important. But that's no excuse for not providing a simple overview written in plain English that might greatly increase the accessibility of your paper. Even ecologists can enjoy a molecular paper, if they are helped to understand it. For example:

> *Overview: To understand which genes are essential for bacterial survival in the legume-Rhizobia symbiosis, we cultured a collection of 10 mutant bacterial strains, each defective in a single gene. We created the 10 mutant strains from a single genotype of a Rhizobium strain that commonly infects soybeans. To represent different stages of the symbiosis we cultured all strains under high (21%) and low (<1%) oxygen. The high-oxygen environment mimics the free-living soil stage prior to root infection. The low-oxygen environment mimics conditions that bacteria encounter once they have entered plant roots. We cultured all strains together, at each oxygen level, for 24 hours and then genotyped the survivors. We replicated each treatment 30 times, and classified genes as essential for growth if strains lacking that gene were not present among the survivors.*

As with our example in the main text, we think this overview works because it explains *why* the study was conducted, and *what* the authors did. The overview is written with non-technical language, that most biologists would be able to understand—even those who are not molecular biologists.

Overviews are also useful in theoretical modelling papers:

> *We examined how the human dispersal rate influences the rate at which a novel virus will spread. We first examined a deliberately simple scenario, where we could derive analytical solutions. In this scenario, we assumed that the dispersal rate is the same across all subpopulations, and that when humans disperse, they are equally likely to go to any of the other subpopulations. We then used an individual-based simulation to explore a more realistic scenario, where we relaxed several of the assumptions that we made in our analytical model. In particular, we examined the consequences of dispersal being more likely to neighbouring groups, and variation in the dispersal rate across populations.*

Again, we think that this explains *why* and *what* was done, with simple non-technical language.

The Details

For some readers, and certainly for reviewers, the overview is simply not enough. Many questions are still left unanswered. After all, the overview only gives us an *impression* of scale, scope, and quality. The devil is always in the detail—but the details don't need to be devilish. So, the rest of your Methods section needs to fill in those details, making it clear exactly *how* you conducted your study. This is the place to include all relevant details of your experimental design, specific protocols, or the details of your mathematical analysis.

The details will almost certainly include multiple subsections, which might have their own subheadings. Your first task is to plan the overall structure of these subsections. In an empirical paper, it's usual to start with a description of your site, system, or study organism. Next, there is often a description of the experimental design or sampling scheme, which might also include a figure. Finally, there are usually one or more paragraphs about your analysis, including the nature of the variables analysed, the types of models fitted, and the assumptions that had to be made.

Let's continue with our ecological example of the impact of nitrogen on grassland plant communities. After reading the overview, several questions come to mind. For example, what was the nitrogen status of the site to start with? How big were the plots? How were the samples taken and how large were they? What exactly were the rates of nitrogen addition? How was the data analysed?

To make sure that we cover all these aspects, we have used four different subsections, each with their own subheadings, but the precise number isn't important. It's only important to divide up the details in a way that makes sense to a reader. Here, the first subsection describes the study site, the second outlines the experimental design, the third gives details of the data collection, and the fourth outlines the analysis:

> *Site: We needed a site with low background levels of nitrogen input, so we chose Castle Hill National Nature Reserve in Sussex, an ancient chalk grassland on shallow sloping soils grazed by sheep. [more details]*
> *Nitrogen treatments: To minimize edge effects and because of high small-scale heterogeneity, we chose a plot size of 3 m × 3 m, much larger than those typically used in other similar*

experiments. We chose three rates of nitrogen addition to bracket the range of values published by Peters (2006), which indicates that chalk grasslands in lowland England typically receive nitrogen inputs ranging from 1 kg per hectare to 5 kg per hectare. The actual rates were ... [more details].

Data collection: In July each year, we sampled three areas measuring 50 cm × 50 cm from within each plot and pooled the resulting biomass. We selected sampling areas by drawing random x and y coordinates with the proviso that the three areas should not overlap. [more details]

Statistical analysis: We analysed how total biomass changed through time as a function of soil nitrogen inputs. We used a linear model with year as a continuous explanatory variable and nitrogen input as a factor with four levels. To meet the assumptions of the analysis total biomass was log-transformed. [more details]

Hopefully, you can see that it's much easier to understand these details because you already had a big picture from reading the overview. The Methods section would feel very different if the overview was missing. The reader would be forced to digest a lot of details about sheep grazing and nitrogen input rates, while still being unclear about the overall nature of the study. They would probably get confused and frustrated, and maybe give up reading your paper.

However, with this structure, they can read the overview and perhaps one or two other subsections that they think are particularly important. We all like to skim papers for the bits that seem important to us, and so we want to make this as easy as possible

Cartoon 3.4 *Lost readers can return to the overview.*

for potential readers. The reader could also return to the overview if they get lost in the details and need a quick reminder (Cartoon 3.4).

Different Details for Different Papers

Different papers can require very different Methods sections, with completely different details. But the same structural suggestions still apply: divide into subsections, which each have their own subheading and purpose.

Let's imagine instead that you carried out a microbial study, examining the genetic basis of antibiotic drug resistance in a bacterial pathogen of humans. In this case, the subheadings for your subsections could be:

Bacterial strains
Genetic manipulations
Resistance assays

Or perhaps you carried out a human politics study, examining how exposure to different types of news item influenced responses on questionnaires about voting preferences. In this case, the subheadings for your subsections could be:

Participants
News manipulation
Questionnaire structure
Statistical analyses

Or perhaps you developed a theoretical model, to examine why cooperation would be favoured in animal societies. In this case, the subheadings could be:

Life cycle
Evolutionary dynamics
Stability analyses

In all cases, you just need to identify the different subsections, and then add some details. This can also make writing less scary. Rather than a blank page for an entire Methods section, you have divided the task into manageable chunks that can be tackled one at a time. Planning your Methods section will make it easier to write.

Finally, think carefully about the order of your subsections. You don't have to present experiments or analyses in the order that you carried them out. Put them in the order that leads to a smoother narrative, and make things easier for the reader to follow.

Scaling Up and Scaling Down

Our above suggestions are for deliberately simple papers. Some papers will be more complicated, with much more work having been done, and so will need many more subsections. In this case, the Methods section might need another level of structure to help the reader navigate their way through.

If the work contains separate parts, then each can have its own subsection, with its own overview. Then, at the start of the Methods, you can still have a general overview to explain the overall structure, in terms of the separate parts. For example:

> *General overview: We first develop a model of plant growth as a function of nitrate availability. We then describe three experiments that were carried out in the lab to test predictions generated by the model. Finally, we describe a larger field experiment, carried out at two sites, which was used to test whether the predictions also held in a real-world setting.*

This overview immediately suggests three main subheadings: *Model*; *Lab Experiments*; *Field Experiment*. The reader also feels a sense of relief—they know what's coming and the text will be broken up into bite-size pieces.

Alternatively, if the end of your Introduction provides a general overview, you might not want to repeat it. Instead, you can just provide a brief overview at the start of each subsection. This might only take a sentence or two. For example:

> *Model: We first developed a model of plant growth as a function of nitrate availability. [Details]*
> *Lab Experiments: To test the predictions of our model, we then carried out three laboratory experiments. [Details]*
> *Field Experiment: To test whether our model predictions also held in a real-world setting, we then conducted a field experiment, across two sites. [Details]*

The Supplementary Information Is Your Friend

When planning the structure of your Methods section, it's also important to think about what *not* to include in your paper. Many journals allow supplementary information (SI), which will only be published online. This has different names at different journals, including supplementary electronic material (SEM), and supplementary online material (SOM). Many people see the supplementary information as somewhere to dump boring-but-useful things, like raw data or simulation code. But it can be so much more than that!

In the main text, you only need to include what is absolutely essential for the paper. So, the supplementary information is the place to include those extra details that some readers want. Possibilities include specific information about how you did something, or alternative ways that you analysed your data. You may also want to include a figure illustrating your experimental design or details of mathematical derivations for a theoretical model. You can even include an additional analysis or piece of related work that is so big that it would disrupt the narrative flow of your main text. Although, remember that most journals require you to refer to each item in the supplementary information within your main text, so you can't dump things in there that are totally irrelevant.

You can also add things to the supplementary information that could never have been included in the main text. For example, you could make a detailed flowchart, or even a video of key methods. Videos can be included either in the supplementary information or at an online site.[1] This kind of material is especially useful if you have developed a new technique that you think other people may want to copy.

In short, cunning use of the supplementary information allows you to get the best of both worlds. You can have a simple and short front end—the main text—that is aimed at general readers, but still include those extra details for an especially interested reader or referee (in the supplementary information). The supplementary information is not just somewhere to hide or dump details in the hope that no one sees them (Cartoon 3.5). Clever use of the supplementary information can help you streamline your paper.

Cartoon 3.5 *The supplementary information is not just for dumping details.*

Referring to Previous Papers

Sometimes you will be following methods that have been fully described in a previous paper. For example, the exact recipe for a medium that is commonly used to grow bacteria. In this case, you could just give a brief overview of the method in the main text, and then say that the details (recipe) are given in that other paper.

But, if you refer to methods in other papers, remember that you can't assume the reader will have actually read that paper, or will go and find it (>95% won't!). So, make sure that you give enough detail that they don't have to read any other papers in order to judge your work.

---

[1] *Jove* is an example of an online repository for videos about methods and concepts (https://www.jove.com).

Filling in the Details

Once you have decided on the overall structure, you need to fill in your structure with clear, simple sentences. The Methods is a good place to start writing because the style should be simple and straightforward. You really don't need to pimp the Methods!

Of course, the order of sentences and their structure still matter. So now it's time to put into practice the core skills that we provided in Chapter 2 (Box 3.2).

Box 3.2 Putting Tips into Practice

It can be easy to fall into the trap of writing up Methods in complicated ways that feel more scientific. Three of the top tips that we gave in chapter two are especially useful for helping to avoid this problem.

1.   Write as you speak

Scientists can come up with amazingly indirect ways of describing what they have done. Below we provide examples and possible alternatives. We highlight the culprit in the first sentence, and a possible alternative in the second:

1. *Due to the fact that* control plants were smaller we had to harvest them earlier.
2. *Because* control plants were smaller we harvested them earlier.

1. *In order to measure* the effect of nitrogen on plant growth, we . . .
2. *To measure* the effect of nitrogen on plant growth, we . . .

1. *We determined the weight of* plants using a microbalance.
2. *We weighed* plants using a microbalance.

1. *We enumerated* the large ruminants in the field.
2. *We counted* the cows in the field.

1. *In order to be able to assess* the aggressiveness of males, *we determined* the number of conflicts that they initiated.
2. *We scored* males for aggressiveness by *counting* the number of conflicts that they initiated.

1. *The county* of Oxfordshire *is located in* south-east England.
2. Oxfordshire *is* in south-east England.

1. We carried out all statistical analyses *in the R statistical environment* (v. 3.3.3, http://www.R-project.org).
2. We carried out all statistical analyses *in R* (v. 3.3.3, http://www.R-project.org).

Avoid archaic words or phrases. Classic culprits are: heretofore, hitherto, henceforth, furthermore, whilst, and notwithstanding. There are always good alternatives and mostly they can simply be left out. Compare:

1. *Henceforth, we* refer to the species by their generic names.
2. *We* refer to the species by their generic names.

2.   Write short sentences, which make one point

Split up different parts of your procedures into different sentences. For example, did you: (1) dry and weigh plants, and (2) analyse them further for chemical characteristics? If so, then this is better written as two sentences. Compare:

1. *After harvesting all plants and drying them in an oven at 80 °C for 24 hours we weighed them on a microbalance to the nearest 0.1 g and then carried out CHN analysis using a combustion analyser to determine the proportions of major elements.*
2. *After harvesting all plants and drying them at 80 °C for 24 hours we weighed them to the nearest 0.1 g. To determine the proportions of major elements we carried out CHN analysis, using a combustion analyser.*

1. *We assumed the likelihood that the bacteria successfully invade a host, f(y), is a positive function of their investment in toxin production (y), given by $f(y) = a+y^b$, and solved for the evolutionarily stable strategy, which is the individual strategy which cannot be outcompeted by any other strategy.*
2. *We assumed the likelihood that the bacteria successfully invade a host, f(y), is a positive function of their investment in toxin production (y), given by $f(y) = a+y^b$. We then solved for the evolutionarily stable strategy, which is the individual strategy which cannot be outcompeted by any other strategy.*

At the same time, don't go too far when dividing up into separate sentences. The odd short sentence is fine, and can even be used to make a strong point. But, too many very short sentences make for difficult reading. For example, compare:

1. *We harvested all plants. We dried the plants in an oven at 80 °C. We weighed the plants on a microbalance.*
2. *We harvested the plants, dried them in an oven at 80 °C, and weighed them using a microbalance.*

It's hard to read the first version out without sounding like a robot, while the second version flows more easily and is still easy to digest. Even though the second version combines three methodological steps, they are all part of one process and not technically challenging. In other words, the point is to measure plant dry weight, and we are just combining the three steps required to do this: harvest, dry, and weigh. Remember that reading what you have written out loud is a great test—most of us naturally speak to the right length, and so reading aloud can alert us to sentences that are too long or short (Cartoon 2.1).

3.   Be precise and consistent

If you have a complicated experimental design or theoretical model it's worth investing some effort thinking about how to name different parts or parameters so that they are intuitively obvious. For example, blocks normally have a specific meaning for experiments, so don't choose to call plots blocks or vice versa!

Avoid ambiguous sentences that can be misinterpreted or are hard to understand. Here are some examples. In each case, try to find a better rephrasing yourself, before reading our suggestion.

> Original: *The Cotswolds and Chilterns are characterized by hilly terrain, higher rainfall, and sedimentary and chalk soils (respectively).*
>
> Diagnosis: It's unclear which characteristics are shared by the two sites and which are unique to each. It's asking too much of the reader to ask them to sort this out.
>
> Alternative: *The Cotswolds have sedimentary soils while the Chilterns have chalk soils; but both areas have hilly terrain and high rainfall.*

> Original: *We focused on two genes: nifH and nifA. They are both essential for nitrogen fixation but have separate promoters and the expression of the former is more sensitive to oxygen levels than the latter.*
>
> Diagnosis: Avoid using 'they' as it's often unclear who or what 'they' are. Also avoid 'former' and 'latter', as this is asking the reader to go back and work out which is which.
>
> Alternative: *We focused on two genes, nifH and nifA, that are both essential for nitrogen fixation. The two genes are independently regulated by different promoters, and the expression of nifH is more sensitive to oxygen levels than nifA.*

> Original: *Three soil cores per site were sampled from 1 m² (where the plant had been previously collected), then treated and analysed.*
>
> Diagnosis: It's unclear whether there is a single sampling location at each site or not. What does the 1 m² refer to? Is it the exact location where the plant had been previously collected?
>
> Alternative: *We sampled soil at all ten sites. At each site, we took three soil cores from a single 1 m × 1 m location where the focal species had been previously collected. The three soil samples were then pooled, treated, and analysed.*

Easy Reading

Methods sections can be dry and tough, so the easier you make it for your reader, the better. As well as the top tips that we gave you in Chapter 2, we suggest four additional tricks to help the reader along.

Put why before how

It's important to include a rationale for each of your choices: for example, why did you add nitrogen in the spring or measure the activity of the *Pnl* gene? The rationale doesn't have to be detailed, but can't be entirely absent. It's also much better if you can place your rationale *before* you give the details of what you did. In other words, put the reason for doing something before the description of how you did it. So, write:

> *To test how chalk grassland responds to increased soil nitrogen, we conducted a field experiment.*
> And *not*:
> *We conducted a field experiment to test how chalk grassland responds to increased soil nitrogen.*

In this very simple case, you might think it doesn't matter much, as both sentences are easy to understand. But it *would* matter if the description was a lot more complicated. For example:

> *The samples were processed by adding 1 ml of a 5M solution of potassium nitrate before cooling for five minutes at 0 °C and then subjecting them to centrifugation at a speed of X mph for 30 minute intervals before pouring off the supernatant, filtering, and adding 3 ml of stain to reveal the presence of the enzyme DNA polymerase.*

We very much expect that when you were halfway through reading this long sentence, you started asking yourself: *but why are they doing this*? Because this is exactly how reviewers feel. And if it's a long, difficult sentence that's hard to understand, reviewers might even start to feel a little angry and frustrated that the authors have subjected them to this experience.

But, by putting the rationale first, this sense of frustration disappears. If you know *why* someone did something, then the details will inevitably make a lot more sense. Let's try again:

> *To reveal the presence of the enzyme DNA polymerase we processed all samples, by adding 1 ml of a 5M solution of potassium nitrate before cooling for five minutes at 0 °C, and then subjecting them to centrifugation at a speed of X mph for 30 minute intervals before pouring off the supernatant, filtering, and adding 3 ml of stain.*

So, help the reader by putting the reason you did something first, not as an afterthought. And never put the reason for doing something in brackets at the end of the sentence (as if it doesn't really matter at all). Here are some further examples. In each case, decide for yourself which you prefer, and why:

1. *We scored each male for injuries on a scale of 1 to 10 (so that we could measure the intensity of fighting).*
2. *We scored each male for injuries on a scale of 1 to 10, so that we could measure the intensity of fighting.*
3. *To measure the intensity of fighting we scored each male for injuries on a scale of 1 to 10.*

And:

1. *We excluded three plants from the analyses (they had been trampled by cows).*
2. *We excluded three plants from the analyses because cows had trampled them.*
3. *Because cows had trampled three plants, we excluded them from the analyses.*

Of course, all rules are there to be broken. Please don't start getting out your red pen and vigorously removing all sentences that put *how* before *why*. There are always going to be exceptions. Like all our advice, it's a good idea most of the time. Not all of the time.

Choose your voice with care

In the Methods section you are describing what you have done, which raises questions about what voice and tense you should use.

The active voice is when the subject of the sentence performs the action. For example:

We weighed the plants on a microbalance.
We added three Daphnia predators to each tube of algal cells.

In a Methods section, this usually means sentences structured: 'we (or I) did something to something'.

In contrast, the passive voice is when the subject has the action performed on them. For example:

Plants were weighed on a microbalance.
Three Daphnia predators were added to each tube of algal cells.

In a Methods section, this usually means sentences structured: 'something had something done to it'.

If you are unsure if a sentence is in the active or passive voice, you can apply the fairy test (Cartoon 3.6).

The active voice is often more logical, clear, and concise. It says something in the way it happened: 'someone did something to something'. Consequently, it often makes things simpler and clearer to the reader. To give some more examples, with passive followed by active, compare:

1. *Whether the presence of herbivores impeded tree growth was experimentally tested.*
2. *We experimentally tested whether the presence of herbivores impedes tree growth.*

Cartoon 3.6 *The fairy test for identifying the passive voice. If you can insert 'by fairies' after the verb then you have passive voice. For example:*
Plants were weighed on a microbalance.
Plants were weighed by fairies on a microbalance.
Alternative versions include the 'by zombies' and the 'by my butler' tests.

1. *Vials were incubated statically for 48 h at 37°C, after which a 2 ml sample was taken.*
2. *We incubated vials statically for 48 h at 37°C, and then took a 2 ml sample.*

1. *The number of plants in each cell was recorded.*
2. *We counted the number of plants in each cell.*

The active voice emphasizes the actor and so makes clear who did what, often in a shorter and more forceful manner. This is especially important in the sections of the paper where you describe what you did and what you found out; that is in the Methods and Results sections.

In contrast, the passive voice can obscure who actually did it, in a way that can cause uneasiness. Who collected your data or developed your mathematical model? Was it you, or was it some fairies that came in the night and did your work for you?

However, don't go rushing away thinking that passive voice is evil and should never be used. Sometimes we want to focus attention on the thing that is affected by the action, and the passive voice can do this. Compare:

1. *We show the strains that we used in Table 1.*
2. *The strains used are shown in Table 1.*

1. *We show how the drug influenced cholesterol levels in figure 3.*
2. *Figure 3 shows how the drug influenced cholesterol levels.*

In both these examples, the second sentence is written in the passive voice. We prefer the passive versions of these sentences because they focus attention on the more important things, the strains and Figure 3, rather than the act of the authors showing them. The passive voice can also be useful for breaking up a long run of sentences that all start 'We'.

To conclude, as a rule of thumb, we suggest that the active voice is the best default option. But, if there is a good reason to use the passive voice for a particular sentence, then do so. All rules can be broken, if it makes things clearer to the reader.

Journals can also have specific requirements. Years ago, it was more common for journals to require the passive voice, because it was felt that it was more subjective. Nowadays, it is generally appreciated that the active voice can be clearer and easier to read, and so journals are more likely to require the active voice. Some disciplines, and some scientists, have been slower to move in this direction, though.

Choose your tense with care

It is normal to use the past tense to describe what you did (what you carried out in the past), but to use the present tense to describe figures (the figures that are here now). So, you might write:

> *We weighed the plants on a microbalance.*
> But:
> *Figure 3 shows how plants respond to additional nitrogen.*

Similarly, in Chapter 4, we will write up the results of statistical analyses in the past tense (because you carried out those analyses in the past).

Use figures and tables

Figures and tables aren't just for the Results section. A figure can be a great way to explain a complicated experimental design, an intricate life cycle, or the assumptions behind a mathematical model. We will discuss how to make figures in Chapter 5.

In an empirical paper, a table can summarize some aspect of how or why data was collected. Tables can also be useful in theoretical papers to summarize the parameters in a mathematical model so that they can be easily referred to, or to hold a collection of equations. Ask yourself: would such tables be useful to the reader? And crucially, would any benefits gained outweigh the cost of making your paper longer?

Details about Details

Hopefully, you now feel equipped to start writing your Methods section. The core principles that we have covered so far apply to most papers, but in the rest of this chapter we look harder at some of the specific details that apply to different types of paper. You can pick and choose which of these to read, depending on the flavour of your science (Cartoon 3.7).

Cartoon 3.7 *What flavour is your science?*

Statistical Methods

Data-driven papers will usually require a subsection describing your analysis. A potential problem is that analyses can get complicated, and the method required for one analysis can depend upon the results of another. For example, depending on whether the data fits assumptions such as normality, you might have to carry out the analyses differently. If we were to write out all the different possibilities, outside the context of the actual data, it could be long-winded and confusing, and make the reader give up.

The solution to this problem is to again divide between overview and details. In the Methods section, just provide an overview of the analyses and the philosophy of your approach. Then in the Results section, alongside the analyses and results, you can provide further details. The overview that you provide in the Methods section can be quite short, terse, and generic. For example:

> *We carried out all statistical analyses in R (v. 3.3.3, http://www.R-project.org). Except where stated, we carried out standard analyses (t-tests and regression analyses) assuming normal errors.*

This example provides an overview of how the analysis was carried out. We have not provided further details on analyses that are so standard we can assume that everyone will know what they are, such as a t-test, least-squares regression, or ANOVA. Some analyses were more complicated and required data transformation, and so that is flagged. To go into more detail in the Methods section would make little sense out of context of the data. Indeed, an equally valid alternative would be to delete the final sentence and leave any mention of normality tests to the Results section. In that case, the Results section would need to provide enough details to make the methodology clear:

> *Results: The species diversity data was not normally distributed (Shapiro–Wilk test: $V = 3.33$, $P < 0.01$, $N = 74$), and so we log-transformed it (Shapiro–Wilk test: $V = 1.30$, $P = 0.26$, $N = 74$). We found that ... [Details].*

We realize that putting 'methods' into the Results section contradicts one of our top tips from Chapter 2, to 'follow the section headings'. But, some blurring of section boundaries can avoid repetition, and make things easier for the reader, so it doesn't contradict our golden rule to put the reader first. In the above example, the excerpt from the Results section describes a result, the problem that arose from that result, and how you fixed it. Not only is it nice to have them together, but splitting the information between the Methods and Results sections would have required more words:

> *Methods: We tested for normality with a Shapiro–Wilk test. In cases where normality was not met, we log-transformed the data, and tested whether this then led to normality.*
> *Results: We found that the species diversity data was not normally distributed (Shapiro–Wilk test: $V = 3.33$, $P < 0.01$, $N = 74$). When we log-transformed the species diversity data it was normally distributed (Shapiro–Wilk test: $V = 1.30$, $P = 0.26$, $N = 74$).*

In Box 3.3, we discuss some other methodological details that could be moved to the Results section. The supplementary information can also provide a very useful space to add analysis details that you think a referee or specialist reader would be interested in, but which would hinder most readers.

Box 3.3 Statistical Details

It is useful to divide statistical methods between an overview and the details. Put the overview in the Methods section, and the details in the Results section. But there are no fixed rules for deciding which information belongs in each section. Try different possibilities, and see what works best for your paper.

For example, to test the robustness of your conclusions, you might have analysed a data set in multiple ways. This could be briefly mentioned in the Methods ('to test the robustness of our conclusions, we carried out several alternative analyses') but then leave the details of those analyses for the Results section. Or you could not mention it at all in the Methods, and just leave everything to the Results section.

An alternative approach is to put the additional analyses in the supplementary information, and just state in the Results section that they led to the same conclusions. Remember that your aim is not to show that you are very clever and have done lots of complicated analyses. Your aim is to convince the reader, as concisely as possible, that your conclusions are sound.

If you have done something unusual, then it may be appropriate to include more details in the Methods. 'Unusual' is from the perspective of the reader, and so will depend upon common practice in the field. So, if you wanted to fix a potential problem by using a particular method which wasn't common procedure, then it could be useful to justify it in the Methods section. For example:

> *Proportion data such as the sex ratio (proportion males) usually have non-normally distributed error variance and unequal sample sizes. To avoid these problems while retaining maximum power, we ... [Details].*

Mathematical Papers

Irrespective of whether you are writing up empirical data or a mathematical model, the general principles outlined above still apply. However, mathematical papers have a few unique problems, which require bespoke solutions. In Box 3.4, we consider how a mathematical paper can be written for readers that lack mathematical skills. In Box 3.5, we give some bonus tips on how to fill out the details of mathematical papers.

Box 3.4 Mathematical Papers: Who Do You Want to Read Your Paper?

Theoreticians face an extra problem, because some of their readers will have a very poor grasp of mathematics. Empirical scientists are lucky because almost everyone reading a scientific paper knows what an experiment is and how experiments can be used to answer questions. In contrast, some empirical workers have relatively limited mathematical skills, and so don't really understand how to construct or use a mathematical model.

Consequently, the first question that a mathematician needs to ask when writing up a paper is whether they also want empirical workers to understand it. In some cases, this might not be necessary. Perhaps you are developing a method for other mathematicians to use. But, such cases will be limited—most of the time, a mathematician will want empirical workers to be able to understand and use their papers. For example, there is little point developing a model of human behaviour if the people collecting data on human behaviour can't understand it.

So, how can you write a mathematical paper that can be understood by those with limited mathematical skills? Once again, the answer is to provide a simple overview, followed by the details. Your overview can be expanded to explain the essence of what you have done, especially your assumptions, in a way that could be understood by all potential readers. This can then be followed by the details of your analysis, aimed at those with mathematical skills. Your Methods are therefore being targeted at two potential audiences: the overview or 'front end' is aimed at all potential readers, while the 'maths details' are just aimed at other mathematicians.

Let us return to the example of a theoretical model developed to examine why cooperation would be favoured in animal societies. In this case, the subheadings and content could be:

Overview: This subsection would contain an overview of the question being asked and the approach you are taking to answer it. It would be aimed at all potential readers.

Life cycle: This subsection would describe the assumptions behind your model, and possibly the logic behind those assumptions. Again, it would be aimed at all potential readers. An empirical worker should be able to assess whether they believe that your assumptions are a good enough approximation to the real world. Most models are only as good as the assumptions they rest on.

Evolutionary dynamics: This subsection would describe technical details of how you analysed your model. It can be aimed at mathematicians—other people who carry out such analyses—and they need to know precise details of your methodology. We are not saying to make this subsection obscure, but rather that certain readers won't be looking at it anyway, so you can include more technical language.

Stability analyses: This subsection would describe more technical details, and again can be aimed at the subset of mathematicians within your audience.

As with an empirical paper, you can also make clever use of an appendix or supplementary information. In the extreme, you could even aim the entire main text at all possible readers, by moving all the analysis details to the supplementary information. The best strategy will depend upon who you are aiming your paper at. Who are the potential readers that you want to attract? How much do you need to balance the needs of different readers?

Box 3.5 Mathematical Papers: Filling Out the Details

There a number of ways in which a mathematical paper can be written to make it more accessible to a wider audience.

Write explanations in plain English first; then follow with the maths

Don't just write maths. It's much better to have a clear explanation in English first.
Imagine that we had produced a theoretical model of toxin production in a bacterial parasite, where the toxin facilitates the infection of hosts. Compare:

1. *We assume that $g = 1 - \alpha y$, where g is the energy available for growth, y is investment into toxin, and α is a constant that determines the cost of toxin production.*

2. *We assume that as the parasite invests more resources in toxin production, the resources invested in growth declines. We capture this trade-off with the relationship $g = 1 - \alpha y$, where g is the energy available for growth, y is investment into toxin, and α is a constant that determines the cost of toxin production.*

The first example probably made your eyebrows furrow and your head hurt. The second example hopefully gave you a better idea of what the mathematician was trying to capture.

Remind readers what your model parameters represent, every time you use them

You have been working with your mathematical model for months, maybe years, and so know it inside out. But a reader will forget what the parameters represent, and so you need to keep reminding them. Always refer to them by their English names, followed by the parameter in brackets. Compare:

1. *We assumed that c was positively correlated with n, by the equation c = bn, where b is a constant.*
2. *We assume that consumption of resources (c) was greater when there were more individuals in the group (n), as described by c = bn, where b is a constant.*

Give parameters simple names

Give parameters simple names with intuitive meanings, as this will help readers understand what they represent. For example, if you need to refer to the number of individuals in a group, call this parameter *group size*, not *microbial population abundance*. If your work builds on previous papers by yourself or others it can be useful to use the same terms and notation, unless the previous work used notation badly. It can be especially confusing to use an existing term with a well-known meaning to mean something different.

Blur section boundaries

The distinction between section boundaries can be less useful in theoretical papers. When you are describing a theoretical model it can be useful to blend sections, by describing the assumptions (Methods), results that emerge from those assumptions (Results), and why those results emerge (Discussion)—all in the same section! In the extreme, if you have distinct analyses, such as two different models, then it can be better to structure your paper around those two analyses rather than traditional 'Methods' and 'Results'. For example, you could describe the assumptions and results of one model, and then do the same for the second model.

Summary

Top Tips for Magnificent Methods
Building Blocks
1. Plan structure
2. Craft a short overview
3. Divide into subsections, with subheadings
4. The supplementary information is your friend
Filling in the Details
1. Apply the specific tips from Chapter 2
2. Put *why* before *how*
3. Choose your voice and tense with care
4. Consider using figures or tables to illustrate complicated issues
Different Tips for Different Papers
1. Consider putting analysis details in the Results section, or the Supplementary Information
2. Write up mathematical modelling so that some of your paper could be read by a reader without mathematical expertise

Box 3.6 presents a summary example. Take the Results quiz before reading Chapter 4 (Box 3.7).

Box 3.6 Summary Example: Time Flies Like an Arrow, Fruit Flies Like a Banana

---

In this box we give two versions of a Methods section. The first has problems, which we have annotated to highlight. This isn't an extreme example—it takes its inspiration from real papers. The second version is improved by the tips in this chapter. Our hypothetical study was examining how fruit flies choose where to lay their eggs (Cartoon 3.8). We have imagined very simple experiments to focus on how it is written rather than technical details.

Cartoon 3.8 *Fruit flies like a banana.*

Version 1
Methods

Insects

The genus Belial belongs to the subfamily Fructusdae, within the Family Huxlinae, which contains over 1,000 species in 60 genera (Burton 2004). The placement of the Belial within the Huxlinae was changed in 2004, following the development of a molecular based phylogeny, which placed them within the Fuctusdae, and not the Tootydae (Burton 2004). This family is found across north, central and south America. The genus Belial contains five species. All species are found from south Mexico to Colombia. B. lys is the most abundant of the Belial species.[2] It causes the largest economic impact.[3] All the Belial species are metallic green, with red eyes, have wings with thickened front edges, and antenna with feathery bristles; the larvae are white (except for black mouth hooks, and brown

---

[2] The Methods section has started with lot of taxonomic details. Do we need these details? And what is being done, or why? The reader will be confused, maybe frustrated. Why do we need to know so much about the taxonomy of these flies?

[3] Possible economic impact seems important. But why? In what way?

abdominal breathing pores).[4] *Belial are fruit flies, which lay their eggs in ripe fruit, which the larvae then consume.*[5]

Cultures

Flies were kept in a 50 cm × 50 cm × 50 cm cage.[6] *Within this cage, the larvae developed for 10 days, before emerging, at 25°C, after eggs had been laid in Burton medium (Burton 1985). The Burton medium*[7] *is made by making a 3% agar medium with apple juice and the addition of 5 g/L yeast extract, and was provided in 5 cm glass jars, with a 3 cm layer of Burton medium, with 10 such jars being placed in the flight cage every two weeks, to maintain a population of flies with overlapping generations.*[8] *A laboratory population of B. lys was used in this study. The cultures were collected in Gamboa during 2004 by Mcgregor et al. (2020). Gamboa is a small*[9] *town located*[10] *in Panama. The population has been maintained in a large population. The temperature of culturing was 25°C. The culture room was humidified. The light: dark cycle was 12: 12 h light: dark.*[11]

Experiment 1[12]

Experiment 1 was carried out in a 20 m × 20 m × 3 m flight cage, in our laboratory facilities. Adult females, which were five days old, and had been mated by males from 2–4 days, were followed in the flight cage, to determine which banana they visited, the time spent on that banana, and to enumerate[13] *the eggs and larvae.*[14] *After the female flies had departed from a chosen host fruit, the observation period was terminated, and they were captured by aspirating into a pooter (OCP supplies). 12 bananas had been placed on the floor of the flight cage.*[15] *At six of these 12 bananas, a mated female was held in location, at the banana, by a 1 cm × 1cm × 1 cm muslin cage*[16] *(supported by a plastic frame).*[17,18]

---

[4] Interesting. Sounds pretty. But did we really need to know this, for the purposes of this study.

[5] This aspect of the basic biology seems key. It would have been useful to put the basic biology earlier, to give context. For example, we can now see how economic impact can arise.

[6] Passive voice. There is a lot of passive voice. Who did the work? Some fairies that came in the night, while the scientists were sleeping?

[7] Burton medium looks an established method. Instead of providing all the details, a citation could just have been given where it was first mentioned.

[8] Look at the size of this sentence. When a sentence is >2 lines, in Word or equivalent, alarm bells should start ringing. This sentence was >3 lines in Word.

[9] Do we need to know how big the town is?

[10] Keep it simple. 'Gamboa, Panama' would have done the job.

[11] There has been a string of very short sentences. It is starting to feel like robot-speak.

[12] This subheading provides no information and doesn't help.

[13] Keep it simple: 'counted'.

[14] Our long sentence alarm is going off again.

[15] At this stage the reader is lost for two reasons: they don't know why both of these experiments have been done, or what exactly has been done. There was no overview / set up, and the crucial parts of the experiment have still to be mentioned. Help!

[16] The word 'cage' is being used to refer to different things, varying from a large flight enclosure to small muslin cages. It is clearer to use different words to avoid confusion.

[17] This sentence provides some key information. But this crucial information is hidden at the end of the description of this experiment. A lot of the earlier text doesn't really make sense until we know the information provided in this sentence.

[18] The replication is unknown. How many females were released?

Experiment 2[19]

Experiment 2 was also carried out in a 20 m × 20 m × 3 m flight cage, at the same laboratory facilities, 17 days after experiment 1.[20] We followed females the same as in experiment 1, so that we could determine which banana they visited, the time spent on that banana, and egg enumerations. As in experiment 1, experiment 2 also involved the placement of 12 bananas on the floor, at regular intervals. At three of these 12 bananas, a mated female was held in location, at the banana, by a 1 cm × 1cm × 1cm muslin cage (supported by a plastic frame).[21] At three other bananas, a virgin female was located in the same way.[22]

Statistical Analyses

We carry out[23] all our statistical analyses in the computing package[24] R (v. 3.3.3, http://www.R-project.org). In the case of experiment 1, we use a number of response variables, including: number of eggs laid, time spent on banana, and banana chosen. The explanatory variable is categorical, whether or not another female had previously been caged at the banana at which observations were taking place, prior to releasing the observation female. As the explanatory variable was categorical, the data is analysed by Analysis of Variance (ANOVA), which compares continuous data between groups, and statistical significance tests used an F-test.[25] In all analyses, we tested for normality with a Shapiro–Wilk test.[26] When normality was not met, as indicated by a significant Shapiro–Wilk test, the data was retested for significance after transforming the data. Possible transformations include log transformation and square rooting. In the case of the time data, normality was significantly not met with untransformed data (Experiment I: Shapiro–Wilk test: V = 3.33, P < 0.01, N = 60; Experiment II: Shapiro–Wilk test: V = 6.66, P < 0.01, N = 60[27]), but was met with log-transformed data (Experiment I: Shapiro–Wilk test: V = 1.30, P > 0.1, N = 60; Experiment II: Shapiro–Wilk test: V = 0.2, P > 0.1, N = 60[28]).

[19] Again, this subheading provides no information and doesn't help.
[20] This paragraph involves a lot of repetition, that could have been avoided.
[21] As above, in experiment 1, the key information is hidden at the end.
[22] Why two types of female—mated and virgin?
[23] Tense—these analyses have already been done.
[24] Need to say 'computing package'?
[25] Lots of unnecessary details have been given about the analyses. These are really simple standard analyses, which need minimal explanation.
[26] It is OK to put things about normality testing here, but this could have been done more briefly. Also, it might have been better to move some into the Results section.
[27] We finally discover the replication, hidden here at the end of the analysis subsection!
[28] It is OK to put normality testing results here, but could have been done more briefly. Also, it might have been better to move into the Results section.

Version 2
Methods

We tested whether females of the fruit fly, Belial lys, prefer to lay eggs on bananas where other females were already present. We carried out two experiments, where we released females into a 20 m × 20 m × 3 m flight enclosure, where they had a choice of bananas on which to lay eggs. In both experiments, we caged single females on 50% of the available bananas. In our first experiment these caged females had already been mated and so were actively laying eggs. To test whether it mattered whether the caged female was laying eggs, we then carried out a second experiment, where 50% of the caged females were mated, and the other 50% were virgins.[29]

Insects

B. lys is a 5–6 mm-long fruit fly,[30] that lays its eggs in a wide range of tropical fruit, which are then consumed by the larvae (Burton 1985). We used a laboratory population of B. lys, which was originally collected in Gamboa, Panama, during May 2004. Since then, a large population has been maintained at 25 °C in a humidified room, at a 12:12 h light:dark cycle (Mcgregor et al. 2020). We maintained the flies in 50 cm × 50 cm × 50 cm cages, providing 10 bottles of Burton medium every two weeks, on which eggs are laid, and larvae could feed (Burton 1985).

*   To obtain adults for our experiments we isolated pupae from our laboratory population in small glass vials (10 mm × 75 mm). We carried out all experiments with females that had emerged five days previously. After emergence, we placed these females with a sugar solution as a food source, and provided them with a same-aged male to mate with between two and four days after their emergence.[31]*

Choice Experiments[32]

We carried out our choice experiments in a 20 m × 20 m flight enclosure.[33] In both experiments, we evenly spaced out 12 bananas on the floor of the enclosure. In both experiments, we alternately caged a female on six of these bananas[34] using a 1 cm × 1 cm muslin cage, supported by a plastic frame (Supplementary Information). In Experiment I, all six caged females were mated. In Experiment II, three of the caged females were mated, and the other three were virgins (50% each type of female). As a control, empty muslin cages were placed at all other bananas.[35]

---

[29] Hopefully this paragraph provides a useful overview. It explains *why* the experiments were carried out, and also *what* was done in the experiments. A reader who didn't care about the details could probably just skip from this to the results.

[30] Explaining up front that it is a fruit fly, and giving an idea of size, so the reader can imagine it. Much more useful to the average reader than taxonomic details.

[31] Even though it was longer, version 1 had missed out how the females were 'set up' for the experiment.

[32] This subheading gives some information, about what kind of experiment, rather than just 'Experiments I & II'.

[33] Instead of describing the two experiments one at a time, in separate subsections, it is more efficient to do both together.

[34] Version 2 is written in the active voice. The sentences have been constructed in a way that prevents the Methods section becoming a huge string of sentences that start 'We ...'. For example, 'In both ...', 'In each experiment ...', and 'To check that ...'.

[35] Even though it was longer, this crucial detail wasn't even mentioned in version 1!

We then released one female at a time into the flight enclosure and followed their behaviour. We recorded when a female landed on a banana, the length of time that they remained there and the number of eggs that they laid. In each experiment, we repeated this with 60 females.[36] *To check that females had been successfully mated, we examined whether their eggs were viable—and in all cases they were (N = 120).*[37] *We followed females until they had laid eggs on a banana and then subsequently moved away from the banana, after which we collected them with a pooter (OCP supplies).*[38] *We recorded the time spent on a banana and the number of eggs laid.*

Statistical Analyses[39]

We carried out all statistical analyses in R (v. 3.3.3, http://www.R-project.org).[40] *Unless stated, we carried out ANOVA, assuming normal errors.*

Comparison

Version 2 is much easier to read, and approximately two-thirds the size of version 1. Despite its smaller size, version 2 provides a clear overview before going into the details, and has even added in some details that were missing from version 1. The text is streamlined in version 2, to avoid repetition and jargon.

---

[36] Replication is given up front.

[37] This is a 'methods check', and so is useful to include here, even though it contains results.

[38] This sentence uses the 'rule of three'—see Chapter 11.

[39] This is a very small subsection. An alternative would be to delete the 'Statistical Analysis' subheading, and just include this text at the end of 'Choice Experiments' subsection.

[40] This statistical analysis subsection is much shorter. It doesn't go into detail on standard procedures, and normality analyses have been left for results.

Box 3.7 Take the Results Quiz before reading Chapter 4

4

Results

By the time a reader reaches your Results section, they should already have a good idea about what you have done (Methods), and why you have done it (Introduction). Now, you finally get to tell them what you have found out. Ultimately, this is the heart of any paper and what it will be judged on.

You may think simply reporting what you found out will be straightforward. But your reader can only take in so much, and it's easy for them to get lost in a black hole of details (Cartoon 4.1). So, how do you make sure that they appreciate your most important results?

We suggest four steps to writing a Results section that can be easily digested by your readers:

1. Plan the structure.
2. Prioritize important results.
3. Use simple language, not statspeak.
4. Blur the boundary with other sections.

Cartoon 4.1 *Don't let your reader get lost in a black hole of details.*

Scientific Papers Made Easy. Stuart West and Lindsay Turnbull, Oxford University Press. © Stuart West and Lindsay Turnbull (2023).
DOI: 10.1093/oso/9780192862785.003.0004

The same principles can be applied, irrespective of whether you carried out an experimental or an observational study, or developed a theoretical model.

Step 1: Plan the Structure

The first step in writing a Results section is to plan out the overall structure. Key questions include:

In what order will you discuss your different results?

Will you use subheadings to organize your results into different subsections, and what will they be?

In what order will you discuss the different points within each subsection?

Follow your Methods section

A good option is to follow the structure of the Methods section. If you carried out two experiments, then your Results section would have a subsection giving the results from the first experiment, and then a subsection giving the results from the second experiment. Depending upon the size of these subsections, they could just be different paragraphs, or you might give them each a subheading.

So, for example, if you carried out three elegantly simple experiments, where the results of each can be explained in one paragraph, then your Results section could just be three paragraphs:

Results
_____

Paragraph 1: Results of experiment A

Paragraph 2: Results of experiment B

Paragraph 3: Results of experiment C

But, if your experiments are more complicated and need more analysis, you might split them up into separate subsections, each with its own subheading:

Results

Experiment A
Paragraph 1: Main results of experiment A
Paragraph 2: Extra analyses from experiment A

Experiment B
Paragraph 3: Main results of experiment B
Paragraph 4: Extra analyses from experiment B

Experiment C
Paragraph 5: Main results of experiment C
Paragraph 6: Extra analyses from experiment C

Follow your questions

Alternatively, it could be better to structure your Results section around questions or hypotheses. For example, if you asked three questions, and used a number of experiments to answer each one, then your structure might look like this:

Results

Question 1
Paragraph 1: Results of experiment A
Paragraph 2: Results of experiment B

Question 2
Paragraph 3: Results of experiment C

Question 3
Paragraph 4: Results of experiment D
Paragraph 5: Results of experiment E
Paragraph 6: Results of experiment F

Or, perhaps you used a single large data set to carry out a number of different analyses, which addressed different questions, in which case a possible structure is:

Results

Paragraph 1: Analyses addressing question 1
Paragraph 2: Analyses addressing question 2
Paragraph 3: Analyses addressing question 3

As in the previous examples, this structure could be expanded if needed. For example, if the analyses addressing one of the questions needed more space, then you could use multiple paragraphs. If all the questions needed multiple paragraphs, then it could be worth adding subheadings.

You can play with the order

Think about the order of your results, and how this could help the reader to understand them. For example, if the results of experiment C help to explain the results of experiment B, then maybe you want to go back and reorder these experiments in your Methods and Results sections, presenting them as A, C, B rather than A, B, C.

Similarly, if you are structuring your Results section around questions, then ask yourself: what is the most useful order in which to answer the questions? If the answers to two questions feed into a third question, then you should present the third one last.

The general point is: you don't have to present your experiments or analyses in the order that you carried them out. Instead, order them in whatever way you think will make it easier for the reader to understand them.

We discuss how to make figures in the next chapter, but authors often make the figures first, to help plan their Results section. Look at the figures together, and think about what order would make them easiest to explain. What order would allow the simplest and clearest narrative flow?

Use subheadings to structure your results

Subheadings lay out the structure of your results, and are a great way to break up a dense Results section. They let the reader take a mental break, and pause, to get ready for something new. A good subheading will also make the content obvious. Good subheadings can take many forms. You could use the title of an experiment:

Predation experiment
Fertilizer experiment

or the topic being examined:

Levels of cooperation
Fertilizer and plant growth

or summarize a main result:

Groups are formed with siblings
Nitrogen fertilizer increased plant growth

You could even use a subheading to outline the question being asked:

Do predators reduce rabbit density?
How are groups formed?
How much do humans cooperate?

Cartoon 4.2 *Subheadings can be recycled from the Methods section.*

Just as the structure of a Results section can follow the structure of the Methods section, subheadings can also be recycled from the Methods section (Cartoon 4.2). For example, suppose that the Methods section contained subheadings that described the experiments:

> *Diet Manipulation*
> *Behavioural Observations*

In this case, you might want to use the same subheadings in the Results section. Consistency helps readers.

You can also use subheadings to provide structure or links between sections. For example, if you carried out two experiments to test the assumptions behind a hypothesis or model, and then one experiment to test theoretical predictions, then you might use:

> *Testing assumptions I: Mate competition*
>
> *Testing assumptions II: Dispersal behaviour*
>
> *Testing predictions: Courtship displays*

You might also use a subheading to justify putting together some analyses that you didn't want to include in other sections. For example:

> *Alternative analyses*
>
> *Robustness analyses*

In short, there are no hard and fast rules for subheadings. Don't be afraid to try different possibilities and see what works best for your paper.

Use the Supplementary Information to streamline your Results

You can move some items out of your Results section into the Supplementary Information. As we discussed in the previous chapter, the Supplementary Information provides a space to put details that are not needed by most readers of your paper, but which some readers might like to see. Possibilities for supplementary material include:

(i)   Tables containing the detailed output from statistical analyses. This could include test statistics and/or parameter estimates.

(ii) Additional figures that were not interesting enough for the main paper, or looked at the data in different ways.

(iii) Alternative analyses and robustness checks. There are often multiple ways to analyse a data set, that are useful to do, but not completely independent. For example, analysing a subset of the data, or analysing the same data set but in a different way. In this case, if they all agree, you might want to put results from the most sensible analysis in the main text, and then all the others in the Supplementary Information. This could be signposted from the main text with a sentence like: 'In the Supplementary Information, we show that we obtained the same qualitative conclusions when we: (1) examined bacteria which were only found in one habitat; (2) reanalysed the data using Smith's correction for sampling bias.'

(iv) Anything that you think the referee or super-interested reader might like to see, or that you really want to include, but which isn't really needed for the main text (or that your co-authors said to take out).

(v)   A script of your analysis, for example an RMarkdown document, that allows someone else to follow its every detail.

The advantage of moving things to the Supplementary Information is that you can shorten the main Results section, which will make things easier for the reader. Remember that *the Supplementary Information is your friend.*

Plan, plan, plan!

The above points emphasize the importance of carefully planning out your Results section before you start writing it. What are your most important results? What order will make your results easiest to understand? Often, the best way to work out what will be best is to plan it out on a piece of paper. List your main points and work out how to order them. Think about where subheadings would help, and whether anything can be moved to the Supplementary Information.

It can also be useful to plan and construct figures before writing. Figures are an integral part of displaying your results, so drawing them can play a vital role in prioritizing what's most important. Some researchers prefer to make their figures before they plan and write their Results section.

A final trick that can help with planning a Results section is to make a small Power-Point presentation of your results. By forcing you to think about how you will present your results to other people, it can become clearer in what order you need to present them, and which are the key results to highlight.

Step 2: Prioritize Important Results

Results sections contain a lot of detail, and you can't expect readers to take it all in. So, what do you most want them to take away from your paper? You can focus your readers' attention by leading with the results that you think are the most important and giving those results more space.

Lead with important results

Within each part of your Results, lead with your most important result, and let the rest follow. For example, imagine that you had done an experiment which showed that eating more chocolate and reducing salt intake had no influence on cholesterol levels, but that drinking more red wine reduced cholesterol. In this case, the positive result with red wine is more important, and the one that you would most want readers to take away from your paper (Cartoon 4.3). Consequently, lead with that result:

> *We found that drinking two glasses of red wine a day reduced cholesterol levels by 30% [analysis details . . .]. In contrast, we found that . . . [other analyses . . .].*

Cartoon 4.3 *Lead with important results.*

This suggestion to lead with important results follows our general writing principle of one main point per paragraph (Chapter 2). In the Results section, this usually means that each paragraph contains only one main result. If there are minor results of lesser importance, they can either be included later in the same paragraph, or given a paragraph of their own.

There are a number of factors that might influence how you choose your most important results:

1. One treatment might have had a large and statistically significant effect, while other treatments had weak or non-significant effects (although be careful to distinguish between the two).

2. One treatment might have been non-significant, but this was in stark contrast to a theoretical expectation or previous results.

3. You may have conducted one primary analysis, and a collection of other analyses that were only there to check the robustness of that result.

4. You may have conducted one main experiment, and then a series of other experiments to check underlying assumptions, or to elucidate mechanisms.

5. Your effect sizes might vary in magnitude or biological importance.

Allocate space according to importance

Another way to focus your readers' attention on the most important results is to allocate space to describing results according to their importance. Or, rather, you should attempt to do this as best you can, given the details of what you must report. We illustrate this principle with an example of where this wasn't done—a results paragraph from an experiment where nitrogen and phosphorous were added to grasslands in a 2 by 2 factorial design. In this experiment, some patches of grassland received nothing, some received nitrogen only, some received phosphorous only, and some received both. The influence of these treatments on both the total plant growth (biomass) and species diversity were measured, and findings reported thus:

> There was no significant interaction term for the influence of adding nitrogen and phosphorous on species diversity ($F_{(1,28)}$ = 0.45, $P > 0.05$). The addition of nitrogen had no significant influence on species diversity ($F_{(1,30)}$ = 1.23, $P > 0.05$). The addition of phosphorous had no influence on species diversity ($F_{(1,30)}$ = 0.89, $P > 0.05$). There was no significant interaction term for the influence of adding nitrogen and phosphorous on plant biomass ($F_{(1,28)}$ = 0.67, $P > 0.05$). The addition of nitrogen led to a significant increase in plant biomass ($F_{(1,30)}$ = 12.34, $P < 0.01$). The addition of phosphorous had no influence on plant biomass ($F_{(1,29)}$ = 0.56, $P > 0.05$).

The above is a fair description of the results, but the significant result—the influence of nitrogen on plant biomass—is a bit lost among all the others. The significant result is both hidden in the middle, and swamped by the other analyses. This can be fixed by refocusing the order (lead with important results), and condensing the space given to some analyses (allocate space according to importance):

> The addition of nitrogen led to a significant increase in plant biomass ($F_{(1,30)}$ = 12.34, $P < 0.01$). In contrast, there was no influence of adding phosphorous on plant biomass nor

any interaction ($F_{(1,29)} = 0.56$, $P > 0.05$; Interaction: $F_{(1,28)} = 0.67$, $P > 0.05$). There was no influence of adding either nitrogen or phosphorous on species diversity (nitrogen: $F_{(1,30)} = 1.23$, $P > 0.05$; phosphorous: $F_{(1,30)} = 0.89$, $P > 0.05$; Interaction: $F_{(1,28)} = 0.45$, $P > 0.05$).

The second version makes the main result stand out and is about two-thirds the size. Double winner! The second version is also easier to read because it is not just a long list of all the results—it starts with a positive result, and then contrasts that with negative results.

Another way to influence the space given to a result, and how likely it is to be noticed, is to use figures to illustrate your most important results. We shall return to figures in the next chapter.

Condense results

One trick that we used above was to condense multiple results into a single sentence. This helps to reduce the length of the Results section. Compare:

1. *The influence of diet supplementation on weight gain did not vary significantly with the sex of the patient ($F_{(1,98)} = 1.23$, $P > 0.05$). The influence of diet supplementation on weight gain did not vary significantly with the age of the patient ($F_{(1,98)} = 2.34$, $P > 0.05$). There was no significant interaction between the sex and age of the patient ($F_{(1,98)} = 0.12$, $P > 0.05$).*

2. *The influence of diet supplementation on weight gain did not vary significantly with the sex or the age of the patient (Sex: $F_{(1,98)} = 1.23$, $P > 0.05$; Age: $F_{(1,98)} = 2.34$, $P > 0.05$; Interaction: $F_{(1,98)} = 0.12$, $P > 0.05$).*

Don't allocate space according to effort

We have emphasized that you should think about what is most important to the reader. You should assume that the reader won't take everything in fully—remember how busy, tired, or stressed they are likely to be (Figure 1.2). Given this problem, what do you want to make sure they notice? If the reader could take only one or two things away from your analyses, what would you want those things to be?

There can be a contrast here with how interesting different parts of the analysis were for you to carry out, or how much effort they required, and how interesting they are to the reader. Perhaps your main and most important result was highly significant and straightforward to analyse, so you aren't quite so interested in it; whereas the bits that challenged you, and took up a lot more time, were a series of complicated analyses that confirmed a negative result. As tempting as it would be to go into detail about your complicated and clever analyses, you probably want to keep these as brief as possible, while focusing on the key results. You could always go into the details of your clever analyses in the Supplementary Information.

To summarize, what matters is the importance to the reader, not the effort you put in. And how much will that reader know or want to know? A typical reader won't remember all the details of your experiment. Some readers will have skipped the Methods section and gone straight from the Introduction to the Results. You need to write a simple, clear Results section, which take this into account.

Step 3: Use Simple Language, not Statspeak

Now that you have your structure planned, how should you fill your different paragraphs and sections? It can be tempting to write up results like the output from a statistical package. However, this approach can make for dry reading that leaves a lot of work for the reader. You can make things easier for your reader by using simple language to explain the direction and magnitude of your results (simple English). This is also a good test of whether you really understand your work. Can you explain your results in a conversational way to your friends?

Include the direction of any effects

Imagine that we had tested the influence of the drug 'Bactokill' on the survival of a bacterial pathogen growing in test tubes. A simple results statement would be:

> *The addition of Bactokill had a significant effect on bacterial growth.*

This may be true, but this not a very useful result. First, we need the statistical evidence for this statement. Often this is most easily done by adding the statistics in brackets at the end of the sentence:

> *The addition of Bactokill had a significant effect on bacteria growth ($F_{3,16} = 127.8; P < 0.001$).*

This sentence now includes a concise summary of the statistical analysis that was carried out (Box 4.1). But, it doesn't tell us what the drug did. We have failed to specify the direction of the effect. In this case it is crucial to know whether the drug increased or decreased bacteria growth! Let's try again:

> *The addition of Bactokill significantly decreased bacteria growth ($F_{3,16} = 127; P < 0.001$).*

Much better! The key point here is that the text describes in simple English what happened, and then the statistical justification for this statement is given in brackets. This puts the focus on what actually happened (the drug reduced bacteria growth), rather than putting the focus on to the statistical test (a significant effect). The lesson is: *use simple language not statspeak* (Cartoon 4.4).
   Compare:

1. *The addition of goji berries had a significant effect on bird visitations to feeders ($t = 6.66$, $n = 30$, $P < 0.01$).*
2. *The addition of goji berries to feeders led to more bird visits ($t = 6.66$, $n = 30$, $P < 0.01$).*

Cartoon 4.4 *Avoid statspeak.*

1. *There was a significant interaction between the levels of nitrogen and phosphorous on plant growth ($F_{1,15}$ = 123.4; P < 0.001).*
2. *The addition of nitrogen led to a greater increase in plant growth, when phosphorous was also added ($F_{1,15}$ = 123.4; P < 0.001).*

Hopefully it is clear that the second alternative—which uses the format: explain in words what happened (and then justify with statistics)—is better in each case.

It is also usual to only include the direction of an effect if it was significant. Compare:

1. *The addition of signalling peptide led to a higher expression of the Roar gene, but this difference was not significant ($F_{1,58}$ = 0.2; P > 0.05).*
2. *The addition of signalling peptide had no significant influence on the expression of the Roar gene ($F_{1,58}$ = 0.2; P > 0.05).*

The first version is longer and also misleading. Noise will mean that different treatments are never likely to be exactly the same, and so it is usual to only describe something as different if it is significantly different.

Provide an estimate of the magnitude of effects

The magnitude of an effect also usually matters (Cartoon 4.5). To use a crude example, most of us would be happy to take a beneficial medicine that also knocked three minutes off our lifespans, even if this was a highly significant difference compared to the control

Cartoon 4.5 *Magnitude matters.*

group. However, we might be much less happy to take a beneficial medicine that was also estimated to knock three years off our lifespans, even if the difference was non-significant (especially if we were concerned that the sample size was rather low or that the data was very noisy). So, don't just report whether an effect is significant. Explain the *magnitude* of the effect and try to give it some context.

Another way to think about this is that the magnitude of the effect puts the emphasis on the subject being studied rather than the statistics *per se*. So, for example, in a biological study, the magnitude of an effect puts the emphasis on the biology and illuminates the 'biological significance'.

Reporting the magnitude of an effect usually means reporting estimates of key model parameters, such as a slope, an intercept, or a difference in the mean value between the control and the treatment group. Reporting estimates like this can make it a lot easier for the reader to grasp the biological significance of your results. Compare:

1. *There was a significant influence of dietary fat content on mouse lifespan ($F_{1,15} = 123.4$; P < 0.001).*
2. *Mice fed a high fat diet had their lifespan reduced by 6.3 (± 2.3) months, compared to mice fed a standard diet ($F_{1,15} = 123.4$; P < 0.001).*

The second alternative provides a much clearer image of what happened. Rather than the nebulous '*there was a significant influence of . . .*', we have an estimate of the magnitude of the effect, and we see its severity. Notice that, in addition to the estimate, a second number is reported in brackets. In this case, it is one standard error. The reader would only know this if you had stated in your Methods section that all estimates are presented with one standard error. The standard error provides a measure of how precise your estimate is. An alternative is to provide the 95% confidence interval. Compare:

1. *Plants receiving additional nutrients were larger than those in the control treatment ($F_{1,35} = 4.5$, $P = 0.02$).*
2. *Plants receiving additional nutrients were 6 g heavier than those in the control treatment (95%CI: 2–8 g; $F_{1,35} = 4.5$, $P = 0.02$). Thus, applying nitrogen at a rate of 1 g per day led to a 20% increase in plant size.*

In alternative 2, we have also elaborated the bare result with a second sentence that tries to give a wider context to the reader, by giving the relative effect as well as the absolute effect.

Not all estimates are means. For example, in regression analysis, the slope of the relationship between two continuous variables is also useful to report. You can use the estimate of the slope to give a 'plain English' explanation of the effect, which gives a reader a sense of its biological importance. Compare:

1. *There was a positive effect of male body size on mating success ($F_{1,35} = 4.5$, $P = 0.02$).*
2. *There was a positive effect of male body size on mating success ($F_{1,35} = 4.5$, $P = 0.02$). The slope of the relationship was 0.56 (95%CI: 0.34–0.78), meaning that the largest males in the population have almost double the mating success of the smallest males.*

In summary, the magnitude of an effect can be described in different ways, and helps the reader to understand the biological significance of your results. If presenting estimates, make sure to include either a standard error or a confidence interval, so that the reader knows something about the precision of your estimates. If you need to present a lot of estimates, then consider putting all or some of them into a table, instead of in the text.

Keep it simple

The Results section offers some dangerous opportunities for 'sounding scientific', and as with all sections you need to stay focused on using simple plain English (Chapter 2). To help you recognize the problem, we provide some real phrases we have encountered (stats omitted for clarity), and a suggestion for how they could be 'translated'.

Original: *Plant reproductive phenology occurred before treatments began.*
Translation: *Plants flowered before treatments began.*

Original: *The addition of soluble nutrients caused plants to achieve a higher final biomass than plants grown with the same volume of water.*
Translation: *The fertilized plants grew larger.*

Original: *Plants experiencing lower rates of pollinator visitation had reduced fecundity in comparison with plants receiving higher rates of visitation.*
Translation: *Plants visited by more pollinators produced more seeds.*

Original: *Following corrections for seasonal variation, the annual trend coefficient of our basic model was significantly negative.*
Translation: *Controlling for seasonal variation, there was an overall decline in insect biomass.*

Mathematical papers

The same principles apply if you are writing up a mathematical model. Explain your results in simple language, not just *mathspeak*. Compare:

1. *This model predicts that $\beta = ad^2$.*

2. *This model predicts that species diversity (β) is correlated with the rate of habitat disturbance (d).*

3. *This model predicts that a higher rate of habitat disturbance (d) will lead to a greater species diversity (β), with $\beta = ad^2$.*

It's better to give the magnitude of an effect, not just whether or not it occurs. Compare:

1. *Our model predicts that the introduction of social distancing measures would reduce the spread of the SARS-CoV-2 virus within the UK.*

2. *Our model predicts that the introduction of social distancing measures would reduce the number of people infected by the SARS-CoV-2 virus in the UK by 80%.*

Box 4.1 Adding Statistics

How much detail should you provide for a statistical analysis? While a detailed discussion of statistics is beyond the scope of this book, we suggest that 'just enough' is the following three pieces of information:

(1) *The test statistic.* This shows the reader which test you did, and its value. Familiar test statistics include t, F, and χ^2.

(2) *The degrees of freedom.* Degrees of freedom are closely related to the sample size. This matters because it will allow a reader to gain an appreciation of the scale of your work and assess whether non-significant results might be due to small sample sizes.

(3) *The P-value.* Your reader may want to know if the result is 'statistically significant' or not.

In our Baktokill example, we provided: '$F_{3,16} = 127.8$; $P < 0.001$'. The test statistic is '$F_{3,16} = 127.8$' and the degrees of freedom are provided as subscripts (F is a ratio of two numbers: the treatment mean square and the error or residual mean square. Both of these terms have their own degrees of freedom, hence there are two numbers separated by a comma. It's the size of the latter that influences the power of the test). The P-value is $P < 0.001$, so this result is statistically significant.

Many statisticians now tell us that we should focus much less on P-values and much more on estimates (which quantify the magnitude of any effects), so provide these too. An estimate (such as a mean, a slope, or an intercept) will tell you the size of the effect, and hence whether it was biologically as well as statistically significant. To show the precision of your estimate, it should always be accompanied by a standard error or a confidence interval.

Putting Different Steps Together

Before going on to the fourth step, we illustrate the interplay between the first three steps with an example in Box 4.2. The purpose of these first three steps is to help clarify the 'story' of your results, with the aim of making the main results stand out as clearly as possible.

Box 4.2 Feeding and Egg Number in Green Tits

The green tit is a small imaginary bird that lives in tree holes. In this study, observations were made at a number of nests, examining both the number of eggs laid by females (termed the 'clutch size'), and how the male and female parents fed their chicks. For comparison, we provide two versions of the Results section from this study (stats omitted for clarity).

Version 1

Results

There was no significant interaction between the sex of the parent and the time of day on the food brought back to the nest (STATS).[1] The rate at which individuals brought food back to the nest did not differ significantly between the male and female parents (STATS). The rate at which individuals brought food back to the nest did not differ significantly depending upon the time of day (STATS). There was no significant interaction between the sex of the parent and the time of day on the food brought back to the nest (STATS). The food brought back to the nest was 90% larvae of the caterpillar Vivarum rufus,[2] 7% other caterpillar larvae and 3% flies[3,4]. The food brought back to nest did not vary significantly between the sexes (STATS). The food brought back to the nest did not vary significantly with the time of day (STATS). There was no significant interaction between the territory size and number of oak trees in the territory of the parent and the number of eggs laid in a nest (clutch size; STATS). There was no significant influence of territory size on the number of eggs laid in a nest (STATS). There was a significant influence of the number of oak trees in a territory on the number of eggs laid in a nest[5] (STATS). There was no significant influence of territory size on the number of eggs laid in a nest (STATS).[6]

[1] Version 1 contains a lot of negative results that could be condensed down.

[2] This caterpillar species is only found on oak trees, where it eats the leaves. This should be pointed out, as it helps explain the clutch size data.

[3] This data on the food brought to the nest is very useful for interpreting the analyses. However, it is currently hidden in the paragraph alongside a list of non-significant results.

[4] It might be better to put the feeding behaviour after the clutch size data. The feeding behaviour data help explain the clutch size data. In addition, it would better represent the time course of events—egg laying (clutch size data) occurred before the chicks were fed (feeding data). However, an argument could be made for the current order, which possibly better reflects causality, because food availability probably determines the optimal clutch size. In cases like this, it's worth trying both and seeing what works best. Whichever order is chosen, the methods should then be put in the same order, so there is continuity between Methods and Results sections.

[5] Sounds interesting, but in what direction? And what was the magnitude?

[6] That was a long paragraph, looking at quite different things. Maybe it would have been useful to split it into two paragraphs?

Version 2
Results

When female green tits held territories that contained more oak trees, they laid a significantly larger number of eggs (STATS). On average, females laid 0.47 (95% CI: 0.31–0.63) extra eggs for each additional oak tree in their territory.[7] In contrast, there was no significant correlation between territory size and the number of eggs laid, so those with larger territories didn't lay more eggs (STATS; Interaction; STATS).

Parents predominantly fed their chicks with larvae of the caterpillar Vivarum rufus, which lives on oak trees. Overall, 90% of food items provided to chicks were V. rufus, 7% other caterpillar larvae, and 3% flies. Both the food brought back to the nest, and the rate at which it was brought back, did not vary significantly depending upon the sex of the parent or the time of day (Food: Sex, STATS; Time, STATS; Interaction, STATS. Rate: Sex, STATS; Time, STATS; Interaction, STATS).[8]

Comparison
Version 2 is an improvement on version 1, provided by using the first three steps for how to produce a digestible Results section. Version 2 has a better structure, prioritizes important results, and uses simple language, not statspeak. These steps lead to a clearer narrative flow, where the key results stand out.

---

[7] We have added in the magnitude to show how much trees mattered. It could be nice to give even more data, such as the mean number of oak trees in territories (and the range). If that data was presented, we could even calculate how much clutch size varied in relation to the density of oak trees.

[8] The text is now short enough that it could potentially be combined into one paragraph. In this version, we chose to divide the different types of data (clutch size versus feeding) between two different paragraphs. But, as ever, there is no hard and fast rule—what would work best for your paper?

Step 4: Blur the Boundary with Other Sections

The Results section is for reporting results—you need to say what you found out. The Results section is not for describing methods or discussing results—there are separate sections to do those things. However, within that guiding principle, there are exceptions, where it can be useful to slightly blur the boundary between paper sections. As ever, the overriding principle is 'do whatever helps the reader'.

Provide reminders

Imagine that you have carried out a large number of experiments, with different aims. Perhaps your first experiments were designed to test the assumptions behind a hypothesis, while others were designed to test the hypothesis itself. You may even have a third set of experiments to discriminate between two competing hypotheses, and now you want to write them up together in a single paper. In this case, even the best of readers could get a bit lost in the analysis, forgetting what the different aims and objectives were—even if they had been laid out clearly in the Methods section. Consequently, it can sometimes be useful to give the reader a brief reminder of the aim or methods in the Results section.

For example, you might want to start one section with: *We carried out two experiments to test our assumption that oranges containing higher levels of vitamin C could be stored for longer without spoiling.* And then start the next section with: *We then tested our hypothesis that the levels of vitamin C in oranges are linked to soil nitrogen content and the temperature on the day flowers were pollinated.*

Reminders should only be added if the benefits outweigh the cost. The cost is a longer Results section. But if reminders can be added concisely, and they are useful to the reader, then the benefit can outweigh the cost. Remember that your reader might not have read your Methods section, or they might be dipping back into this section of the paper, after initially reading it a long time ago. The key word is 'concisely'—you just need a brief reminder, not an overly long repetition of sections that rightly belong in the Introduction or Methods. And if you do add reminders, remember to *put why before how* (Box 4.3).

Box 4.3 Write Why Before How

In Chapter 3 we showed how it could help the reader by writing *why* before *how*, to provide context when writing Methods. This trick is just as useful in the Results section. If you remind the reader why you carried out an analysis, then it will help emphasize why the result is important. For example:

In support of our hypothesis that additional nutrients would enhance plant growth, plants were larger in the nutrient treatment than in the control ($F_{(1,148)} = 34.5, P < 0.01$).

In contrast to Bloggs et al. (2016), we found a strong positive correlation between phosphate concentration and algal biomass ($F_{(1,33)} = 45.6, P < 0.01$).

In support of Rowan's (1999) theory of genome size, there was a significant positive correlation between genome size and the number of cell types found in an organism ($F_{(1,48)} = 54.3, P < 0.01$).

Explain your theory

When presenting the results of a mathematical model, it can be useful to explain why that result was obtained, and not just present the naked result (Cartoon 4.6). Strictly speaking, explaining a result counts as 'discussion'. But, it can be much clearer and more efficient to do this as you go along, rather than leaving all explanation to the Discussion section, when a reader is quite likely to have forgotten a lot of your results. It also allows the reader to fully understand one part of the modelling before they move to the next, and can reduce repetition between the Results and Discussion sections. So, give a prediction, and then spend a sentence or two explaining that prediction. For example:

Our model predicts that Bluehead wrasse should mature as females, and then change sex from female to male when they reach approximately 80% of their maximum adult body size (Figure 3). This prediction arises because the benefit from increased body size is larger for males than for females (Figure 2). Females of all sizes are able to breed, but larger females are able to produce more eggs, and so fitness increases relatively linearly with body size in females (Figure 2a). In contrast, only the largest males are able to successfully defend harems of females, and hence gain mates. Consequently, fitness increases exponentially with body size in males (Figure 2b). Sex change is favoured at the body size where the relative fitness of males becomes greater than the relative fitness of females (Figure 3).

Cartoon 4.6 *Don't just present naked results.*

One way to structure the results of a theory paper is to give each prediction its own paragraph, in which you present the prediction and then explain it, before moving on to the next prediction in a new paragraph.

Investigate journal requirements and options

Some journals require a combined 'Results and Discussion'. If so, it can often be useful to discuss each result as it is presented. Depending on the details of what you did, results and discussion could be combined in one paragraph, or divided between several paragraphs. So, in the case of our fictional paper with three different types of experiment, if the results and explanation were super simple and concise, you might end up with:

Results & Discussion
_____

Paragraph 1: results of experiment A; discussion of experiment A
Paragraph 2: results of experiment B; discussion of experiment B
Paragraph 3: results of experiment C; discussion of experiment C

But, if things were more complicated, and needed more explanation, you would need separate subsections with multiple paragraphs.

Results & Discussion

Experiment A

Paragraph 1: Main results of experiment A
Paragraph 2: Extra analyses from experiment A
Paragraph 3: Discussion of experiment A

Experiment B

Paragraph 4: Main results of experiment B
Paragraph 5: Extra analyses from experiment B
Paragraph 6: Discussion of experiment B

Experiment C

Paragraph 7: Main results of experiment C
Paragraph 8: Extra analyses from experiment C
Paragraph 9: Discussion of experiment C

With a very complicated paper, even this structure might not be enough. Perhaps some or all subsections need at least one or more discussion paragraphs. This will depend upon the details of your paper—there are no hard and fast rules; you need to step back and think about what is required.

Some journals allow you to choose between separate or combined Results and Discussion sections. If so, you need to consider what would work best for your reader. For example, if you have multiple complicated analyses, then it could be useful to discuss each as you go along, and so you could opt for a single combined section. Alternatively, if your results really don't need much discussion, then you could keep your sections separate. Plan out different options, and see which you think would work best.

Summary

Four Steps to Digestible Results
Step 1: Plan the structure
1. Follow your Methods or questions
2. Play with order
3. Use subheadings to structure your results
4. Use the Supplementary Information to streamline your results
5. Plan, plan, plan!
Step 2: Prioritize important results
1. Lead with important results
2. Allocate space according to importance

3. Use figures to emphasize results

4. Condense results

5. Don't allocate space according to effort

Step 3: Use simple language not statspeak

1. Include the direction of any effects

2. Provide an estimate of the magnitude of effects

3. Keep it simple

4. Use simple language, not just mathspeak

Step 4: Blur the boundary with other sections

1. Provide reminders

2. Explain your theory

3. Investigate journal requirements and options

Box 4.4 presents a summary example.

Box 4.4 Summary Example

In this box we give two versions of a Results section, for the fruit fly study that we introduced in Box 3.6. The first version isn't great, and we have annotated it to highlight potential problems. The second version is improved, following the steps given in this chapter.

Version 1

Results

Very little is known about how fruit fly females choose where to lay eggs, and especially whether females aggregate to lay eggs, or avoid other females.[9] The oviposition[10] sites chosen by females is of paramount importance to agriculture, because it determines the damage that fruit flies will cause to fruit crops. Fruit flies are attracted to a wide variety of fruits, which has considerable financial implications for farmers, because they cause considerable damage.[11] Little attention has been paid to whether females aggregate to lay eggs on certain fruit, or whether they avoid fruit where other females are ovipositing. We examined this question. Our long-term aim is to understand the behaviour of females to be able to better reduce the impact on fruit crops. Traps can be used to lure females away from crops, and stop them ovipositing.

[9] This paragraph is Introduction, not Results. Although a reminder can be useful, we are only describing a couple of relatively simple experiments, and so this reminder is too long.

[10] Jargon, not plain English. 'Oviposition' is 'eggs laying'. In addition, the language in this paragraph is not consistent—it switches between 'egg laying' and 'oviposition'.

[11] This sentence has the cause (damage) before the effect (implications).

Experiment 1

The time spent at a banana was log-transformed because it was not normally distributed[12] *(Shapiro–Wilk test: V = 3.33, P < 0.01; Figure 1a)—after transformation, it was normal (Shapiro–Wilk test: V = 1.30, P = 0.26; Figure 1b*[13]*). The amount of time that females spent feeding on a banana did not vary depending upon whether there was another female there*[14] *(F$_{(1,58)}$ = 0.13, P > 0.05; Figure 2*[15]*). The accumulated rate at which females oviposited on bananas depended significantly*[16] *upon whether another female had been caged there*[17] *(F$_{(1,58)}$ = 13.4, P < 0.01; Figure 3). The fruit chosen by females for oviposition depended significantly*[18] *upon whether another female had been caged there*[19] *($\chi^2_{(1)}$ = 23.4, P < 0.01, N = 60; Figure 4). The time spent at a banana*[20] *depended significantly*[21] *upon whether another female had been caged there*[22] *(F$_{(1,58)}$ = 12.3, P < 0.01; Figure 5).*

Experiment 2

Female[23] *flies may adjust their foraging behaviour in response to a wide range of factors.*[24] *Our result that females adjusted their behaviour significantly, in response to the presence of other females, suggests that allowing for these behavioural motivations will be an important avenue in the development of lures for future research.*[25] *In our second experiment, we confirmed the results of our first experiment.*[26] *The time spent at a banana was log-transformed because it was not normally distributed (Shapiro–Wilk test: V = 3.67, P < 0.01; Figure 6a)—after transformation, it was normal (Shapiro–Wilk test: V = 1.67, P = 0.33; Figure 6b). The amount of time that females spent feeding on a banana did not vary depending upon whether there was another female there (F$_{(1,58)}$ = 0.89, P > 0.05; Figure 7). The accumulated rate at which females oviposited on bananas depended significantly upon whether another female had been caged there (F$_{(1,58)}$ = 16.7, P < 0.01; Figure 8).*

[12] This is a tough start to the paragraph! Is it the most important result to lead with? Also how is described before why—the transformation (log-transformed) is given before the reason why transformed (not normal). Also written in passive voice.

[13] This seems a waste of a figure, to show how transformation influenced normality. Remember space according to importance (and figures take space).

[14] The results are not ordered well. One possibility would be to order results as they happened—e.g. fruit choice, and then what females did at the fruit. Another possibility is to focus first on the significant differences.

[15] This version of the Results appears to show everything as a figure. There has been no attempt to work out what to show and what not to show. No thought about what to prioritize / give more space. No thought about what order, or how to group results. The structure was not planned carefully.

[16] Written in statspeak—doesn't tell us direction or magnitude.

[17] Passive voice.

[18] Written in statspeak—doesn't tell us direction or magnitude.

[19] Passive voice.

[20] This result has been separated from the fact that the data for this analysis had to be transformed to make it normal.

[21] Written in statspeak—doesn't tell us direction or magnitude.

[22] Passive voice.

[23] Look at the length of this paragraph. Alarm bells should be ringing—it looks too long. Very indigestible.

[24] This sentence is vague Introduction material; not helpful.

[25] There are lots of problems with this sentence. It is discussion, not results. It is vague, statspeak, too long.

[26] This sentence, and the next six sentences, are a very lengthy repetition of the same pattern found in the other experiment. The same wording is even used as with the first experiment, suggesting a cut and paste. This section could be written a lot more succinctly.

The fruit chosen by females for oviposition depended significantly upon whether another female had been caged there ($\chi^2_{(1)}$ = 18.9, P < 0.01, N = 60; Figure 9). The time spent at a banana depended significantly upon whether another female had been caged there ($F_{(1,58)}$ = 14.5, P < 0.01; Figure 10). In contrast, there was not a significant influence on the total number of eggs laid by a female, depending upon whether the other female[27] was mated or a virgin[28] ($F_{(1,58)}$ = 1.45, P > 0.05; Figure 11). The time feeding on a fruit[29] did not depend upon female status[30] ($F_{(1,58)}$ = 1.67, P > 0.05; Figure 12). The time spent at a fruit did also not vary significantly depending upon female status ($F_{(1,58)}$ = 1.23, P > 0.05; Figure 13). The likelihood of visitation[31] showed a non-significant difference[32] depending upon female status ($\chi^2_{(1)}$ = 0.12, P > 0.05; Figure 14[33]).

Version 2
Results

In our first experiment, when we caged another female at a banana, we found that females were more likely to visit that banana, and they then remained longer and laid more eggs.[34] Females were 4.5 times[35] more likely to visit[36] a banana with a caged female than bananas without a caged female[37] (95% CI: 3.2–5.8; $\chi^2_{(1)}$ = 23.4, P < 0.01, N = 60; Figure 1). Females stayed at bananas with a caged female 1.3 times as long, and laid 1.5 times as many eggs, compared with bananas without a caged female (time visited: 95% CI = 0.9–1.7, $F_{(1,58)}$ = 12.3, P < 0.01; Eggs: 95% CI = 1.2–1.7, $F_{(1,58)}$ = 13.4, P < 0.01; Figure 2[38]). The time spent at a banana was not normally distributed (Shapiro–Wilk test: V = 3.33, P < 0.01), and so we log-transformed it prior to analysis (Shapiro–Wilk test: V = 1.30, P = 0.26). In contrast, whether there was another female caged at a banana did not influence the amount of time that females spent feeding on that banana ($F_{(1,58)}$ = 0.13, P > 0.05; Figure 3[39]).

[27] Vague—what 'other female'?
[28] This is a clunky sentence. The effect is written before the cause. Also, it is not clear which previous result the 'In contrast' refers to.
[29] Lack of consistency. Sometimes 'fruit' is used, sometimes 'banana'.
[30] Vague—what aspect of female status?
[31] Not simple English.
[32] What does a 'non-significant difference' mean? Two measures of something are incredibly unlikely to be exactly the same. The question is whether they are significantly different, and the magnitude of that difference.
[33] The number of figures has got crazy.
[34] This first sentence gives a gentle overview, before going into the details. Note that cause (experimental caging of female) is given before effect (on behaviour).
[35] Here and below an estimate of magnitude is provided (4.5 times). You can see how much the factors really mattered. Confidence intervals of this estimate could also have been given.
[36] The results are structured (ordered) by time, first describing banana choice, and then what flies did when there at the banana.
[37] When providing a comparison, like '4.5 times more likely', you need to specify what you are comparing it to.
[38] These two behaviours at the banana can be put together in one figure (time there and eggs laid).
[39] This figure shows a negative result, and so a figure was not necessarily needed (to prioritize important results). However, the result contrasted with previous work and so was included.

In[40] *our second experiment, we found that the behaviour of females was not influenced by whether the female caged with the banana was mated or a virgin. There was no significant influence of whether the caged female was mated or a virgin on the likelihood that females visited a banana, the time that they spent there, the number of eggs that they laid, or the time that they spent feeding on the banana*[41] *(likelihood visited: $\chi^2_{(1)}$ = 0.12, P > 0.05; time visited: $F_{(1,58)}$ = 1.23, P > 0.05; eggs: $F_{(1,58)}$ = 1.45, P > 0.05; time feeding: $F_{(1,58)}$ = 1.67, P > 0.05). As in our first experiment, the time spent at a banana was not normally distributed (Shapiro–Wilk test: V = 3.67, P < 0.01), and so we log-transformed it (Shapiro–Wilk test: V = 1.67, P = 0.33). As in our first experiment, we again found that when we caged another female at a banana: (a) females were more likely to visit that banana, remained longer, and laid more eggs (likelihood visited: $\chi^2_{(1)}$ = 18.9, P < 0.01, N = 60; time visited: $F_{(1,58)}$ = 14.5, P < 0.01; eggs: $F_{(1,58)}$ = 16.7, P < 0.01; time feeding: $F_{(1,58)}$ = 1.03, P > 0.05); and (b) there was no significant influence on the amount of time spent feeding on the banana ($F_{(1,58)}$ = 0.89, P > 0.05).*

Comparison

Version 2 is about three-fifths the size of version 1, but it gives more information. It describes the direction and magnitude of effects, making it much clearer what happened in the experiments. Results are described in simple language, making it much easier to imagine or conceptualize the results. Version 2 also emphasizes the more important (positive) results.

Version 2 was produced by cutting material that was less useful. We removed any text that was not 'results', and we used fewer figures. The reader will be able to get through version 2 much more quickly, and so they are less likely to get bored and go off and do something else instead. And because it is shorter and better focused, the main results are laid out more explicitly and clearly, so that the reader will be more likely to understand and remember them.

We shall provide figures to go with these results in the next chapter.

---

[40] Notice that version 2 did not use subheadings. It is just a small results section, and so no need. But subheadings could be added. If subheadings were used, they should be more informative than just 'Experiment 1' and 'Experiment 2'.

[41] All these non-significant results have been lumped together into one sentence.

5

Figures

Figures can be an incredibly efficient way to communicate large amounts of information. They also allow a reader to understand even complicated data or concepts with relatively little effort. So, given that we are trying to make things easy for our hypothetical reader, you won't be surprised to learn that we think good figures are essential to any paper.

However, making figures inevitably raises several questions. These include:

How do you choose which results should be illustrated with a figure?

What is the right level of complexity?

How should figures be labelled?

How much detail is required in a figure legend?

We answer these questions by breaking figure construction down into three simple steps:

1. Choose your figures carefully.
2. Make simple, effective figures.
3. Edit to achieve graphical excellence.

We then provide basic principles for each step. We illustrate our points with examples based on simulated data. Our aim is to provide a brief introduction on how to choose and design figures, rather than an exhaustive treatment. There is a large literature, beyond the scope of this book, on 'data visualization', which combines statistical analysis, graphic design, and psychology.[1] We also discuss how similar questions arise when designing tables, and consider when a table should be used instead of a figure.

---

[1] Two excellent entry points into that literature are: *The Visual Display of Quantitative Information* by Edward Tufte (2001), and *Fundamentals of Data Visualization: A Primer on Making Informative and Compelling Figures* by Claus Wilke (2019).

Scientific Papers Made Easy. Stuart West and Lindsay Turnbull, Oxford University Press. © Stuart West and Lindsay Turnbull (2023).
DOI: 10.1093/oso/9780192862785.003.0005

Step 1: Choose Figures Carefully

Figures, like posters or movie trailers, do not need to tell the whole story. Instead, they are used to emphasize and clarify the most important points. The good news is that you are the expert on your research, so you get to decide what's important. Use figures to show the messages that you want the reader to take from the paper and to help them understand your results in a simple, friendly way.

Prioritize key results

When writing your Results section we emphasized the importance of giving space to results in proportion to their importance (Chapter 4). Careful choice of figures can help you to achieve this goal. By illustrating only the key results—the ones that you really want readers to notice—you will put the focus on what is important.

As with writing, it's important to be concise. If you make too many figures, then the important ones will be hidden among the less important, reducing their impact. Consider what the main points are that you want illustrate or explain, and how figures can help. Constructing more figures can help to get across more points, but it will also dilute your message. So, only make figures for the key results and don't try to illustrate every finding. A good rule of thumb is to step back and sketch out two to six figures that will communicate the story of your paper.

For example, imagine that we were choosing figures for the green tit analyses in Box 4.2 (Page 61). The key result was that female green tits with territories that contained more oak trees laid significantly more eggs, and this needs illustrating (Figure 5.1). Although there was a second important result—that parents predominantly fed their chicks with larvae of the moth *Vivarum rufus*—this can be presented in the text, and

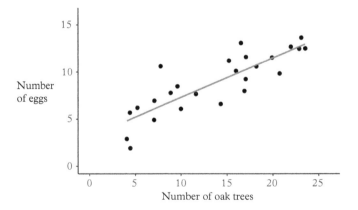

Figure 5.1 *Female green tits with more oak trees in their territory laid more eggs. Each data point represents one female, and the fitted line is the best fit linear model.*

doesn't really need a figure (although we shall return to this later). There were also several non-significant results that were important but probably don't need to be illustrated. The advantage of having just one figure from multiple analyses is that it focuses the attention of the reader on the most important result.

Illustrate complicated results

Some results or patterns are particularly hard to describe in words. This might be because the data is complicated, or because you are doing something unusual with the data that many readers might not have encountered before. Figures can provide a great solution to both these problems.

Statistics are often about making comparisons, but these comparisons can be complicated. Perhaps you had to control for multiple factors in your analysis or you might have had several variables of interest. A figure can be used to clarify these comparisons by showing subgroups, or by clarifying which comparisons have been made. Figure 5.2 provides an example where the statistical analysis was a paired t-test, and the pattern only becomes clear when the pairing structure in the data is included in the figure. Panel (b) is more informative (clarifying the pattern that females are larger) and more faithful to the analysis (which was a paired comparison).

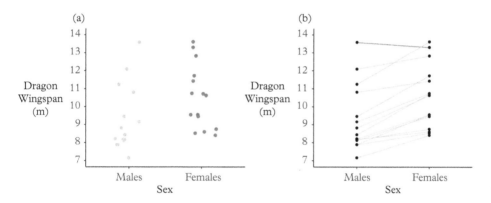

Figure 5.2 *A comparison of male and female wingspan across 15 dragon species. (a) and (b) show the same data, but the wingspan of males and females of the same species are connected in panel (b), reflecting the statistical comparison made.*

Remember that you will have spent a lot of time thinking about your analyses. Consequently, even complicated results might seem clear to you, but it's important to step back and think about how a reader might engage with your findings. Try explaining your findings to a friend who doesn't know your work very well. What did they find especially complicated? Did making a drawing help them to understand? If the answer is yes, then perhaps that diagram needs to be included in your paper.

Don't just illustrate results

Remember that figures aren't only used for showing results. You can use figures in other parts of your paper, such as: (1) illustrating a hypothesis or competing predictions in the Introduction (Chapter 6); (2) drawing out a complicated experimental design or mathematical model in the Methods (Chapter 3); or (3) using a figure to explain a key concept that the rest of the paper depends upon (Introduction or Methods). For example, if your paper was about the different ways in which genes can be transferred between bacterial cells, then you could use a figure in the Introduction to illustrate the various possible mechanisms (Figure 5.3). Figures like this won't always be needed, and should be used sparingly, but they can really help.

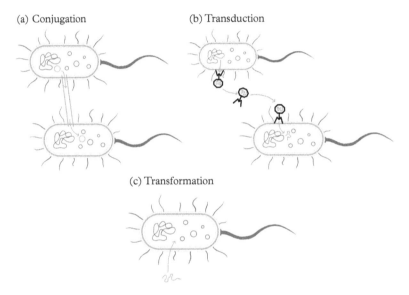

Figure 5.3 *Types of horizontal gene transfer in bacteria. (a) Conjugation: a conjugative plasmid (blue circle) moves through a pilus into the cytoplasm of the recipient cell. (b) Transduction: the bacterio-phage genome (blue) inserts itself into the bacterial genome (purple), where it replicates and is repackaged into the phage capsule (black). Sections of bacterial DNA (blue) can be inadvertently incorporated into the capsule along with the phage DNA, so providing a mechanism for bacterial DNA to be transferred to other cells. (c) Transformation: a bacterial cell enters a state called competence where it can take up free DNA (blue) from the environment.*

Step 2: Make Simple, Effective Figures

A good figure will communicate information simply and efficiently. It should allow the reader to obtain maximal understanding with minimal effort. We suggest eight guiding principles to making simple, effective figures.

Choose what to communicate

Each figure should only illustrate a single point or a small number of related points. If you try to illustrate too many points with one figure then you will make things harder for the reader and they might miss the main point entirely. Actively choose which points would benefit from illustration and which points don't really need it.

For example, in Figure 5.4, we provide an expanded version of Figure 5.1, which also includes information on prey items and territory size. The cost of this additional information is that it distracts from the main result. In this case we think that the cost outweighs the benefit and so prefer Figure 5.1, but perhaps you disagree? It can be helpful to try different options and see what works best.

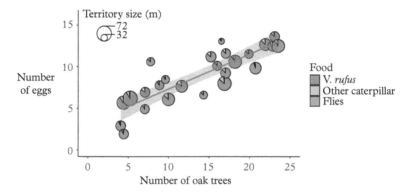

Figure 5.4 *The number of eggs laid by female green tits (clutch size) versus the number of oak trees in a female's territory. Each data point represents one female, and the line is the best fit linear model. The size of each data point shows the territory size for that female, and the colours in the pie chart of each data point show the proportion of different food items collected by that female.*

Show the data

Figures are sometimes produced which include information from the analysis, such as a fitted line, but be careful when considering whether to ditch the data entirely. The amount of data might be prohibitively large; but any analysis is only as good as the data on which it is based. Even plotting means and intervals can make it impossible to see the shape of the raw data, including outliers. So, if you can include the raw data in your figures, do so.

Figure 5.5 shows the relationship between whisky consumption and the temperature of the fire produced by different species of dragons. The brave scientist who carried out this study fitted a linear model with a quadratic term and then plotted the fitted line both with and without the raw data. If only the fitted line is shown (a), then there appears to be a clear domed or humped relationship, but when the raw data is included (b), the hump shape is not strong—indeed, an asymptotic function could be more appropriate.

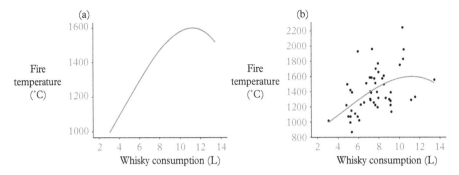

Figure 5.5 *The temperature of a Dragon's fire increases with whisky consumption (litres per day). (a) shows the fitted line but omits the raw data; while (b) includes both. (a) also uses a reduced range on the y-axis range, that does not cover the full range of the data.*

Another important reason for plotting the data is to make the sample size clear. If it's not possible to include the raw data, then the sample size could also be included in the legend, together with other essential details of your experimental design or analysis.

Make figures as simple as possible

Make figures that are as simple as possible. Designing simple figures involves making the right choices about the types of figures to produce and which to avoid. For example, avoid three-dimensional figures, as most people can't process them, while over-cluttered figures, or figures with too many colours, will strain the eyes of the reader. Compare the panels in Figure 5.6. In the 3-D version (a), the relationships are obscured. Panels (b) and (c) show that there is a clear positive relationship between red wine consumption and cholesterol level but panel (c) is simpler and more faithful to the analysis (the number of glasses of wine consumed was treated as a continuous variable). Other problems with the box plots in (b) include that they obscure the raw data, and most people don't know what the various lines actually show, so this information would need to be provided in the legend.

Simple, effective figures are more likely to be remembered by the reader, and hence increase the chance that your paper will have impact. While drawing your figures, consider the following questions: could my figures be used in an undergraduate lecture or a textbook? Could my friends and family understand this figure, with just a bit of help?

Remember, we aren't saying that you should never make complicated figures. Some data or concepts are complicated, and the figures will probably have to reflect this complexity. Rather, you should make your figures as simple as possible, given what you are trying to illustrate.

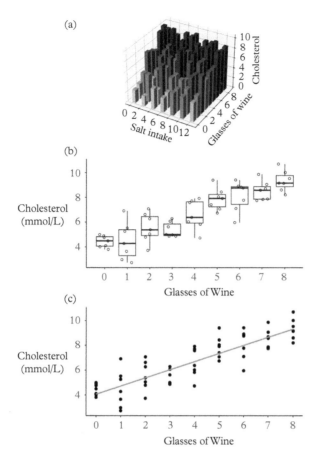

Figure 5.6 *Relationships between blood cholesterol level and different aspects of diet. (a) the relationship between cholesterol and both wine and salt intake (per day). (b) the relationship between cholesterol and wine consumption per day with a plot shown for each level of wine consumption. (c) the relationship between cholesterol and wine consumption per day with individual data points plotted separately.*

Show the result

Figures should show the result as clearly as possible. Make sure that the reader doesn't have to decipher the data before they can see or appreciate the result. If a reader has to carry out multiple calculations in their head, and then put them together for themselves, they will get confused and frustrated.

Compare the panels in Figure 5.7. Panel (a) includes too much information so the reader must do quite a lot of work if they want to extract the take-home message. In con-

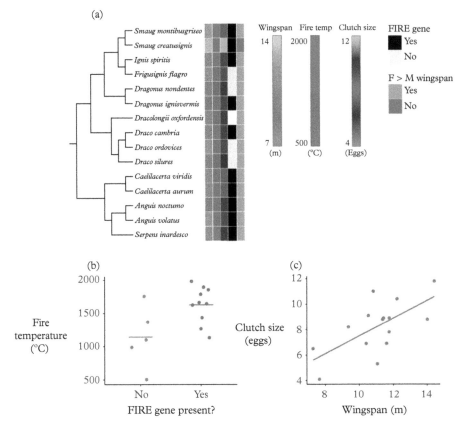

Figure 5.7 *Dragon life-history*. *(a) a phylogeny of 15 dragon species, with details of their life-histories. (b) dragon species carrying the FIRE gene breathe a higher temperature fire than species that do not. (c) dragon species with a larger wingspan lay more eggs per breeding attempt (clutch size).*

trast, panels (b) and (c) show two key patterns in much simpler ways. Including error bars can also be an important part of showing the data and the result (Box 5.1).

Box 5.1 Error Bars

If your figures display estimates (e.g. mean values in different treatments), then including an appropriate error bar is essential. However, there are different types of error bars and it's very important to make clear which you are using. Variability in the data is often quantified using the standard deviation, which is a measure of dispersion. However, if you want to indicate the precision of an estimate, then you need to provide a standard error or a confidence interval.

These quantities belong to the estimate, so we talk about the *standard error of the mean*, or the *confidence interval of the intercept or slope*.

In some circumstances, the confidence interval has the additional benefit of allowing the reader to judge the significance of a statistical test. For example, if the 95% confidence interval of a mean does not include a particular hypothesized value, then we know that mean is significantly different from that value at $P < 0.05$. Similarly, if two 95% confidence intervals do not overlap, then we can conclude that the means are significantly different at $P < 0.05$. Of course, if they *do* overlap, then we can't be certain that they are *not* significantly different and we must look at the outcome of the relevant test. However, this property of confidence intervals is useful and means that they are often preferable to showing one standard error.

Communicate truthfully

It is of paramount importance that a figure doesn't mislead, either deliberately or otherwise. For example, compare the panels in Figure 5.8. In panel (a) the *y*-axis has not been extended to zero, and it appears that the drug is having a large effect. In contrast, panel (b) makes clear that the drug is having a very small effect. Although, some would say that most of the space in panel (b) is devoted to non-useful information, so another solution is to draw attention to the break in the *y*-axis in panel (a) by using a squiggly line.

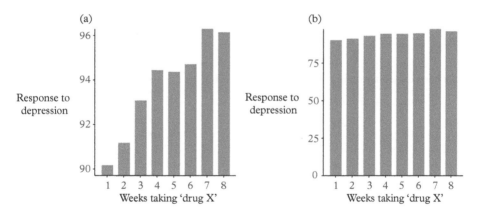

Figure 5.8 *The effect of taking the experimental drug Can-d for different numbers of weeks on a clinical trial for depression. The two panels show the same data, but with a different range on the y-axis.*

Label clearly and use annotations sparingly

Label and annotate your figures to make them as easy as possible for readers to understand. In Box 5.2 we provide some tips for labelling axes. In Box 5.3 we make suggestions for annotations, including keys, colours, and other labels. The best way to label and annotate can depend enormously on the type of figures you are constructing, and we don't have the space here to be exhaustive. Our general point is to consider different options and make active choices—don't just use the default options of your graphical software!

Write simple and clear legends

Legends should be written using simple language, rather than jargon defined elsewhere in the paper. Only include information that is essential to understand the figure—legends are not a place to sneak in extra details. As with elsewhere in the paper, be concise. The legends in this chapter provide examples that you can adapt and use.

Construct stand-alone figures

A reader should be able to understand a figure without consulting the main text. Some readers might just be skimming through your paper, or even ignoring the main text and just looking at the figures. Consequently, your figures could be your only chance to explain your results and get your reader interested. Readers may also return to a paper that they have read before to remind themselves of the crucial points. If you can convey these with your figures, then your work is more likely to have impact. Really good figures might mean that the reader barely has to read much of your paper at all!

Box 5.2 Labelling Axes

It can be easy to add labels that are obvious to you, but won't make sense to the reader, who hasn't been buried in the research for months. Our top four tips are:

1. Use intuitive and short axis labels

Label your axes with intuitive words, like 'Population growth rate', 'Cognitive load', or 'Genome size'. Avoid abbreviating labels with symbols or acronyms, such as 'λ', 'CL', or 'C-value'. If labels are not easy to understand, then the reader will have to keep jumping between the figure and the legend. The more easily a reader can understand your figures, the more likely they are to keep reading your paper—and they are more likely to cite it too. Compare the panels in Figure 5.9, which both show the same data. Panel (b) can be understood without reading the legend, whereas panel (a) cannot.

Avoid overly long axis labels. Just a few words should do. You can always specify in the legend precisely what you mean—for example, 'in this paper we use genome size to mean the amount of DNA contained in a haploid nucleus'.

As with many aspects of figures, it's useful to try different options. If adding a few extra words provides necessary clarity, then it could be worth a longer label. For example, in Figure 5.9b, we used relatively long labels on the *x*-axis because this enabled us to describe the treatments in full. This makes it easier for the reader to glance at the figure and understand the result without consulting the legend.

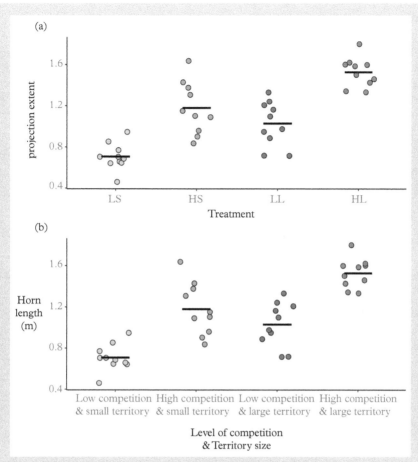

Figure 5.9 *Rearing conditions and unicorn horn length. (a) Unicorn horn length (projection extent; m) for unicorns reared under low (L) or high (H) competition, and on small (S) or large (L) territories. (b) Unicorn horn length on territories of different quality. Each data point represents one territory.*

2. Include units with your axis labels

You should usually include units in your axis labels (compare the *y*-axis labels in panels (a) and (b) of Figure 5.9). Similarly, in theoretical papers, the coefficient for a term in a model should be included in the axis label of a figure. For example, use 'Selection coefficient (*s*)' or 'Growth rate (*g*)'.

3. Write axis labels horizontally

Labels on the *y*-axis are often written vertically, parallel to the axis. However, while vertical labels can save space, writing the label horizontally makes it easier to read. Compare the *y*-axis labels in panels (a) and (b) of Figure 5.9.

4.   Make labels legible

Check the font size! Text can sometimes look different in a statistical software package than in the exported graphic. Make sure that the final submitted figures have legends and labels that will be large enough to read comfortably in the printed paper. Nobody wants to pick up a magnifying glass to read your figures! See Box 5.5 for some examples of badly sized labels. Remember that different readers might go through your paper in different ways, including on a computer, tablet, phone, or even in hard copy. You might also want to produce different versions of your figures with different label sizes to use in a presentation.

It can take time to work out the right combination of font size, line thickness, and so on. Consequently, it can be worth working this out in the statistical or mathematical software that you use, and then keeping the same code as a template to reuse every time you need to make a new figure. You might even want multiple templates for different purposes.

Box 5.3 Annotation and Graphical Design

Cunning use of annotations, colour, and other aspects of graphical design can greatly aid the reader. *Keep it simple and keep it clear.* Our top tips are:

Include a key to symbols within the plotting area

The inclusion of a simple key that denotes the meaning of different symbols or line types means your figure can be decoded more easily. As with axis labels, use short and intuitive labels. In Figure 5.10 this key can be included within the plotting region. If there isn't a natural space within the plotting region, then consider putting the key to one side, as long as it doesn't over-clutter the figure.

Figure 5.10 involves some redundancy, because both colour and symbol shape are used to distinguish the high and low soil quality treatments. This redundancy can be useful because it gives an insurance against possibilities such as the reader printing in black and white, or problems distinguishing colour combinations.

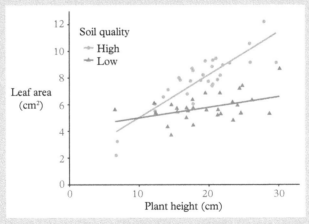

Figure 5.10 *The relationship between leaf area and plant height, for plants grown in high and low soil quality.*

Annotate to clarify results.

It can be useful to annotate a figure to indicate the results of an analysis. For example, if your data are categorical and you are interested in differences between the mean values in different treatments, then a horizontal line and an asterisk (⋆) can be used to signify when groups are significantly different (Version 2: Figure 2 in Box 5.5). Although post hoc comparisons have their supporters and their critics.

Use a colour-blind-friendly palette.

Approximately 5% of people have colour vision deficiency, which makes it hard for them to distinguish certain colour combinations, with red and green being the most common example. Consequently, if you produced a figure that relied upon red/green to distinguish two different types of data point, then some readers would miss this distinction. Many statistical and mathematical packages have solved this problem, either directly or through add-ons, with colour-blind-friendly palettes, for example by using combinations such as blue/orange. Another option is to vary shape as well as colour, as we have done in Figure 5.10.

Shape your graphics

Humans appear to be better at interpreting graphs when they are framed within certain shapes. In particular, a rectangle stretched horizontally is better than a square (compare panels (a) and (b) in Figure 5.11). A good rule of thumb is to use a rectangle that is roughly 50% wider than it is tall.

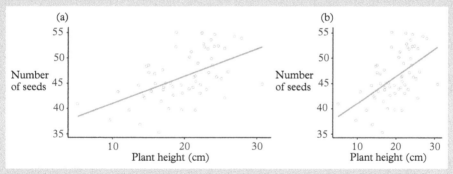

Figure 5.11 *Taller plants produce more seeds. The different panels show the same data and the same regression line, but the plotting region is rectangular in (a) and square in (b).*

Label graphical regions directly

If you have carried out theoretical work, use intuitive labels for different regions of the graph. For example, in a study that looks at the effect of environmental factors on competitive outcomes between species A and B, label the regions: 'A wins' or 'B wins' (see Figure 5.12). This is a much better solution than shading regions in different ways and then providing a translation in the legend below. If it doesn't result in an over-complicated figure, it might be possible to indicate qualitative and quantitative outcomes within the same plot. For example, use a heatmap to indicate the size of an effect, and lines (clines) to separate areas with different qualitative outcomes.

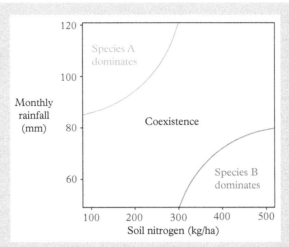

Figure 5.12 *The predicted outcome of the model under different rainfall conditions and amounts of soil nitrogen. With high monthly rainfall and low soil nitrogen, species A dominates. With low monthly rainfall and high soil nitrogen, species B dominates. In intermediate conditions, the two species coexist.*

Choose a helpful title

The first sentence of a figure legend should be a brief title. In a single-panel figure, you could summarize the result, for example *Supplementary feeding increased hedgehog survival* (see also Figures 5.1 and 5.5). Alternatively, you could use a more neutral title that simply states the relationship shown, such as *The impact of supplementary feeding on hedgehog survival* (see Figures 5.2 and 5.4 for other examples). In figures with multiple panels, you probably need a more general title, for example *Feeding experiments* or *Factors affecting hedgehog survival* (see Figures 5.6 and 5.7 for other examples). Try different options and think about what works best. To help them stand out, you can also write figure titles in bold type (e.g. Figures 5.3, 5.7, 5.9, and 5.11)

Be consistent

Be consistent across panels and figures in your use of symbols, colour, etc. For example, if you use green to indicate something in one figure, then use it in the same way in other figures. Plan your colour scheme early on, and then apply it across all figures.

Step 3: Edit to Achieve Graphical Excellence

Figures can always be improved by editing. In Chapter 11 we provide some strategies for editing text. While many of the same strategies can be applied to figures, there are six tricks that are especially useful when editing figures.

Leave and forget

Leave the figure and come back to it a few days later. Try to put yourself in the place of someone who is new to the data. Is the figure clear and easy to understand? Is the figure simple and uncluttered? Are the axes and labels the best options? How well can you understand the figure without reading the legend?

Maximize information content

A good figure should draw the reader's attention to the take-home message, so edit your figure to focus on the data. One way to do this is to think carefully about how the ink in a figure is used. Edward Tufte,[2] who is something of a legend when it comes to data representation, showed that figures can often be improved by reducing ink used for anything other than data (non-data ink).

Removing 'non-data ink' is often a good way to reduce clutter and help the data stand out. Forensically analyse your figure and work out what could be removed. For example, try removing gridlines, frames and unnecessary shading. We provide an example in Figure 5.13. The data is much clearer in panel (b), where the non-data ink is minimized. Another example is provided by Figure 5.6, where there is less non-data ink in panel (c), compared with panel (b).

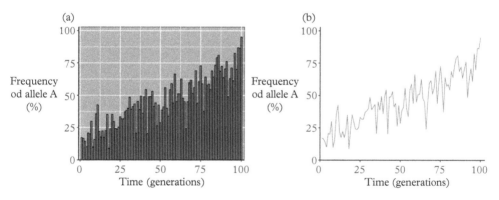

Figure 5.13 *Frequency of allele A in an evolving population. Both panels show the same data, but plotted in different ways. Panel (b) has reduced the non-data ink.*

As always, don't be too rigid. It is possible to go too far, so try different options to determine what you think makes the data stand out best. There are also occasions where it can be useful to add gridlines, for example when particular values have special meaning. In Figure 5.13, the value of 100% is of special significance because the allele

---

[2] Edward R. Tufte, *The Visual Display of Quantitative Information* (2001), Graphics Press LLC.

has reached fixation. Consequently, it could be useful to keep a single horizontal line at 100%, perhaps with the word fixation written on it.

Remove fluff

It can sometimes be tempting to make figures optically interesting. However, rather than help the reader, these attempts often obscure the data, and should generally be removed. Your aim is to make something easy to understand, not something that is artistically interesting. For example, compare the fluffed (a) and simple (b) versions in Figure 5.14. Fluff is often non-data ink.

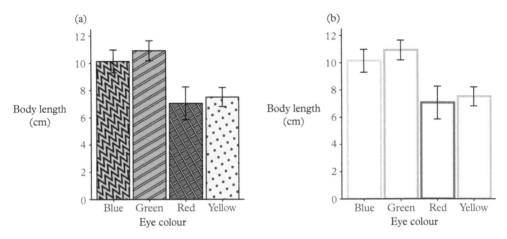

Figure 5.14 *The mean body length of rainbow-eyed snakes with different eye colours. Error bars are 95% confidence intervals. The two panels show the same data. Version (a) has been touched up (fluffed) to make it more 'interesting'. Version (b) has been simplified to focus on the data.*

Reduce reliance on the legend

A reader should be able to grasp the purpose and main message of a figure without reading the full legend. With more complicated data, the reader will probably need to consult the legend, but you should still try to make it as easy as possible and keep text in the legend to a minimum. Examine your figures critically and think how reliance on the legend could be reduced. Again think of a reader who is just looking through or looking back at your paper.

Try different options

There is no single correct method to produce a great figure. The best option can depend on the type of data, the form of analysis that has been carried out, and even the target audience, so don't be afraid to experiment and see what works best for each figure. This could involve tweaking a figure, or even thinking about different approaches (Box 5.4).

The principles that we have outlined can even pull in different directions because figures vary enormously in their form and complexity. Sometimes a super-simple figure that barely requires a legend will be possible. But at other times, the best option will be a more complex figure, with multiple panels, where the reader must refer to the legend. Similarly, you will sometimes be able to strip a figure to the bare minimum of ink, but at other times you will need to include more details to save the reader from having to keep looking things up in the legend.

Look at what others do

Look at the figures in papers from your field to get inspiration. Some problems that you are grappling with might already have been solved, or a certain type of data might usually be presented in a particular way, that everyone is used to. In addition, there are numerous more detailed and targeted guides on making figures. These include dedicated websites, papers, and books, aimed at both broad and specific audiences.[3]

Box 5.4 Try Something Different

It's easy to keep making the same style of figures, both within and across your papers. That might be because you've hit on the best possible approach, but it's sometimes worth stepping back and thinking about how you might change your figures—or even do something totally different.

1.   **Consider adding a title**

The inclusion of a title can be a good idea. This can be applied either to a single figure or to the different components of a multi-panel figure. Sadly, some journals won't allow titles, so check the instructions for authors.

2.   **Consider using multiple panels**

Consider whether it's better to combine multiple graphics within the same figure. For example, returning to the data in Figure 5.1, if two different factors had influenced the number of eggs laid by female green tits, then we could have expanded the figure to include two panels. Adding extra panels allows you to show more data, but be careful not to overdo it. Don't be tempted to increase the number of your figures by stealth, by adding lots of extra panels!

Using multiple panels within one figure can also be a good way to link up results. For example, different results from the same data set, or from the same experiment, can be

---

[3] *The Visual Display of Quantitative Information* by Edward Tufte (2001) provides an excellent overview, which is expanded upon in Tufte's other books. *Fundamentals of Data Visualization: A Primer on Making Informative and Compelling Figures* by Claus Wilke provides a detailed guide to all aspects of figure construction. *The Analysis of Biological Data* by Michael Whitlock and Dolph Schluter is an example of a book aimed at a more specific audience. Both general and field-specific advice can be found on the Web by searching with terms such as 'better' or 'effective' figures.

grouped together within a single figure (see Version 2: Figure 2 in Box 5.5). Such clustering can provide a theme, and save the reader from having to skip between different figures to get the whole story.

3.   Consider adding fancy touches

Consider whether you can do anything to make a point more clearly, or to bring out an additional subtlety from the data. There is no simple rule here, but rather our suggestion is to step back and think about what could work in a particular case.

For example, if we again return to the green tit data from Box 4.2, we note that 90% of food items provided to chicks were caterpillars of *V. rufus*, 7% were caterpillars belonging to other species, and 3% were flies (Diptera). This result could be summarized very simply within the text. But if it was a particularly important result that was worth highlighting, then we could consider how an appropriate figure might be drawn. One obvious option would be a pie chart (Figure 5.15a), but it isn't very inspiring. An alternative is to have drawings of the different organisms, with the area of each drawing proportional to the representation in the birds' diet (Figure 5.15b). The second option is certainly more striking and more likely to be remembered. Another option would be to use a technique called 'treemapping', where different parts of the data would be represented as nested rectangles.

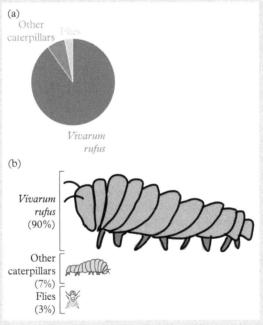

Figure 5.15 *Food items brought back to the nest by green tits. The different panels show the same data, summarized by either a pie chart (a) or by the area taken up by the drawing (b).*

The possibilities for adding fancy touches will depend on the kind of data that you have. For example, if a figure showed data from different species, you could try using silhouettes of species rather than standard data points.[4] Similarly, silhouettes or photos of species could be added onto tips of a phylogeny.

4.   **Consider making a high-information graphic**

The human eye is remarkably good at fine-scale discrimination. Consequently, a lot of information can be included in one figure. This could be large amounts of data in a single figure or many different small panels. When multiple panels are used to show a series of figures, with the same combination of variables, they are termed 'small multiples'. We provide an example in Figure 5.16.

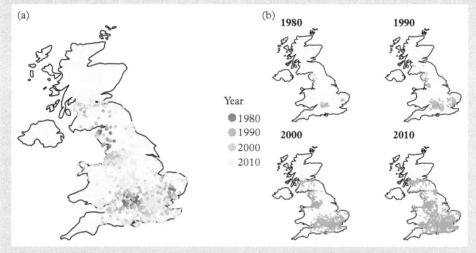

Figure 5.16 *The increasing range of the green tit within the UK. Both panels show the same data, but presented either as a single figure (a) or several small multiples (b).*

---

[4] A handy resource for organism silhouettes is provided by http://phylopic.org.

Tables

Tables can be used to summarize statistical analyses or parameter estimates. As with figures, the aim is to make things as easy as possible for your reader, by maximizing the information, in a simple and easy-to-understand format. Consequently, analogous points to those we have given for figure design also arise for table design.

Tables have several possible advantages. A table can allow a lot of detailed information to be summarized in a way that can be easily understood. For example, a table can show

Table 5.1 *The percentage of different food items brought back to the nest by green tits. Percentages are calculated from 3,228 food items, over a four-week period, at 40 nests.*

Food	Percentage of total food
Vivarum rufus	90%
Other caterpillars	7%
Flies	3%

how the data is divided up between different classes, or the statistical details of several analyses. Putting data in a table can have more impact than using a simple bar graph or pie chart and can allow other researchers to use that data more easily. For example, let's return again to the data collected on the food that green tits bring back to the nest (Box 4.2; Figure 5.15). This data could be presented in a table rather than in a figure:

The relative benefit of using a table will depend upon the data, and how you want it to impact upon readers. In the specific case of the green tit data, we prefer Figure 5.15b to Table 5.1, but it does seem excessive to dedicate a whole figure to this data. A cunning trick would be to place Figure 5.15b as an inset within Figure 5.1. This would allow us to include Figure 5.15b without taking up too much extra space.

Using a table for statistical details also has advantages and disadvantages. A table can help to unclutter the main text, and make it easier for a reader to grasp the details of certain types of more complicated analysis (e.g. when comparing the output of several alternative models). A potential disadvantage of using a table is that it can take up much more space. In Chapter 4 we suggested putting all the analyses at the end of sentences—this approach leads to relatively uncluttered text, without taking up the space of a table. Consequently, given our general aim to keep a paper as short as possible, it is often better to just report the outcome of statistical tests in the text and put additional details of analyses in the supplementary material. The supplementary information can be an excellent location for tables that summarize additional details of analyses, or alternate ways of doing analyses.

Another key consideration when choosing whether to add a table is how much space and emphasis you want to give to results. Tables make results stand out. They can also be useful for summarizing the parameters used in a theoretical paper, so the reader can easily find what each of your terms denotes. The total number of tables and figures is sometimes limited by journals, in which case the addition of a table needs to be traded off against other possibilities, such as losing a figure.

If you do include a table, apply all the same principles that we have suggested for making figures. Make sure your table stands alone and can be understood without reference to the text. Just as with figures, readers will often jump to tables without reading the text. Make sure that the labels in the table are simple and clear, and that the legend gives all necessary information.

Summary

Simple Steps for Fantastic Figures
Step 1: Choose figures carefully
1. Prioritize key results
2. Illustrate complicated results
3. Don't just illustrate results
Step 2: Make simple, effective figures
1. Choose what to communicate
2. Show the data
3. Make figures as simple as possible
4. Show the result
5. Communicate truthfully
6. Label clearly and use annotations sparingly
7. Write simple and clear legends
8. Construct stand-alone figures
Step 3: Edit to achieve graphical excellence
1. Leave and forget
2. Maximize information content
3. Remove fluff
4. Reduce reliance on the legend
5. Try different options
6. Look at what others do

Box 5.5 presents a summary example. Take the Introduction quiz before reading Chapter 6 (Box 5.6).

Box 5.5 Summary Example

In this box we provide figures for the fruit fly study that we introduced in Box 3.6. We provide two versions of the figures for that study. We first provide a bad version of the figures, annotated to highlight potential problems (version 1). We then provide an improved version of figures, following the advice given in this chapter (version 2).

We decided that writing bad figure legends to accompany the bad figures in version 1 would be too confusing. Consequently, the legends in version 1 just briefly say what each figure shows. In version 2 we provide examples of full legends.

The figures in this box stand alone, in the context of this chapter. However, they also correspond to the Results sections for the fruit fly study, that we provided in Box 4.4. Specifically, the two bad versions go together (version 1 in Box 4.4 with version 1 in this box), and the two good versions go together (version 2 in Box 4.4 with version 2 in this box).

Version 1

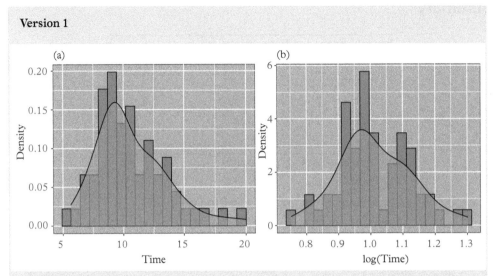

Version 1: Figure 1. *(a) The distribution of time that flies spent on bananas is non-normal. (b) Log transformation produces a distribution that is closer to a normal distribution. Blue bars display a histogram of the data (y-axis not shown). The green area is a density plot of the data.*[5]

---

[5] Problems with Version 1: Figure 1 includes: (i) bad choice of what to include as a figure, let alone have as the first figure—this normality test could just be in the text, without a figure (or added to supplementary information); (ii) axis labels are too small; (iii) redundancy of histogram and density plot used together; (iv) time has no units, and could refer to either 'time spent on banana' or 'time feeding'; (v) panels could have more useful title than just 'a' and 'b'; (vi) 'density' (on y-axis) is not explained; (vii) lots of non-data ink; (viii) the y-axis label is written vertically, not horizontally.

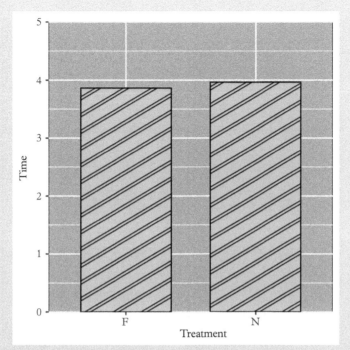

Version 1: Figure 2. *Difference in the time flies spent feeding on bananas on which a female was caged (Treatment 'F') compared to bananas on which no female was caged (Treatment 'N').*[6]

[6] Problems with Version 1: Figure 2 include: (i) poor choice for second graph given negative result; (ii) pattern within bars serves no purpose beyond aesthetic (i.e. 'fluff'); (iii) time has no units and is not clear whether this is 'time on banana' or 'time feeding'; (iv) the *x*-axis label 'Treatment' is uninformative; (v) *x*-axis labels are uninformative (Treatments F and N)- there is plenty of space to use the full words; (vi) the *y*-axis has wasteful empty space at the top; (vi) tiny axis labels; (vii) the data is shown as a bar graph (not raw data) with no estimate of the variation (e.g. 95% confidence interval or standard error); (viii) the background shading is distracting non-data ink; (ix) the *y*-axis label is written vertically, not horizontally.

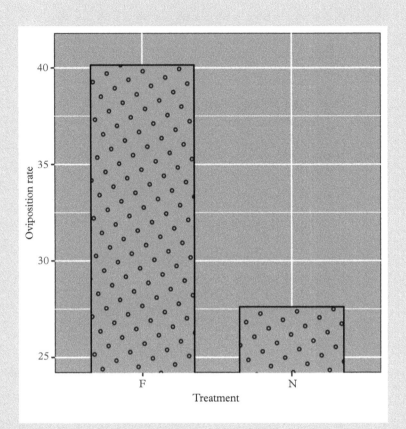

Version 1: Figure 3. *Flies laid more eggs on bananas on which a female was caged (Treatment 'F') compared to bananas on which no female was caged (Treatment 'N').*[7]

[7] Problems with Version 1: Figure 3 include: (i) potentially misleading *y*-axis scale—starting at 24 makes the difference between bars look much bigger; (ii) the *y*-axis label of 'Oviposition rate' is jargon for number of eggs laid; (iii) *x*-axis labels are uninformative (Treatment, F and N); (iv) tiny axis labels; (v) no error bars or confidence intervals; (vi) the *y*-axis label is written vertically, not horizontally; (vii) both the bar shading and the background shading are distracting non-data ink that serve no purpose (except to make the reader feel a bit queasy);

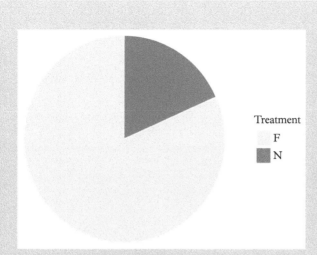

Version 1: Figure 4. *Flies were 4.5 times as likely to visit bananas on which a female was caged (Treatment 'F', yellow area) compared to bananas on which no female was caged (Treatment 'N', blue area).*[8]

[8] Problems with Version1: Figure 4 include: (i) use of pie chart gives no idea of sample size, just the average proportions; (ii) inset legend gives no useful info—could have just labelled sections directly with 'F' and 'N' (or something more informative!); (iii) so much non-data ink for essentially two data points.

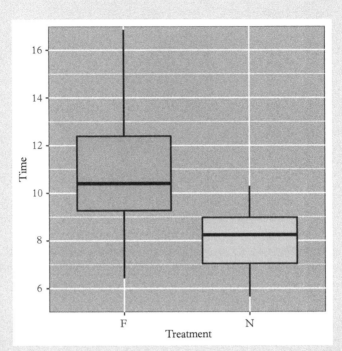

Version 1: Figure 5. *Flies visiting a banana on which a female was caged stayed longer compared to flies visiting a banana on which no female was caged. 'F' stands for 'female caged on banana' (orange box). 'N' stands for 'no female caged on banana' (blue box). The middle line in each box is the median, the upper line is the 75th percentile, and the lower line is the 25th percentile. The vertical 'whiskers' display the range of data points outside the 25th to 75th percentiles, excluding any outlying data points more than 1.5 × the height of the box.*[9]

[9] Problems with Version 1: Figure 5 include: (i) this is one of the most important results, but it is hidden as the fifth figure; (ii) the box plot is an improvement on the simple bars used in Figures 2 and 3, but takes a lot to explain what it means, and outliers have been excluded—would have been simpler to just show the data; (iii) the *y*-axis doesn't start at 0; (iv) the *x*-axis labels are uninformative and small (Treatment, F and N); (v) the *y*-axis label is written vertically, not horizontally; (vi) lots of non-data ink; (vii) the difference in colour across the two treatments is not consistent with earlier figures which also compared across bananas with and without females.

Version 1: Figures 6–13. Figures 1–5 above are the figures to go with the 'Experiment 1' subsection of the Results in 'Version 1' of Box 4.4. We have not included the other figures cited in that Results section (Figures 6–13), because the problems would become repetitive.[10] Instead, we jump to Version 2.

Version 2

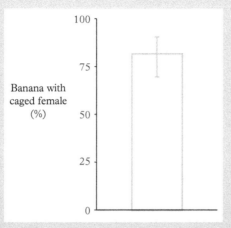

Version 2: Figure 1. *Females prefer to visit bananas where another female was caged. The bar shows the percentage of females that visited a banana where another female was caged. The error bar is the 95% confidence interval.*[11]

---

[10] The fact that we are excluding Figures 6–13 also emphasizes that Version 1 of the Results section in Box 4.4 suggested far too many figures! There has been no effort made to prioritize and focus on important results—instead every possible figure has been included. A reader could get lost and bored in all these figures.

[11] A simple figure, but including it helps emphasize this result. *y*-axis label is horizontal. *y*-axis label is a bit long, but allows full explanation without reading legend. Low amount of non-data ink. Legend is short, but includes title and summary sentence. Compare with Version 1: Figure 4.

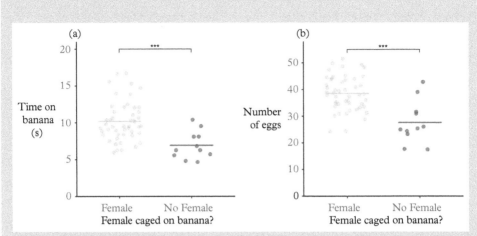

Version 2: Figure 2. *Females both stayed longer and laid more eggs when there was another female caged at the banana.* Each data point is one of the 60 females released, and the horizontal lines show the means (***=P < 0.001[12]).[13]

<hr />

[12] The version 2 figures have been annotated when there are significant differences between treatments, with a horizontal line and an asterisks (*)—although post-hoc comparisons have their supporters and their critics!

[13] Shows raw data and means. Colours are consistent with Version 2: Figure 1. Labels are horizontal, a good size, and informative (no need to read legend). Legend is short, but includes title and summary sentence. Compare with Version 1: Figures 3 and 5.

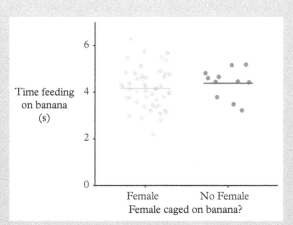

Version 2: Figure 3. *Time spent feeding on bananas is independent of the presence of other females.* Each data point is one of the 60 females released, and the horizontal lines show the means.[14]

Comparison

Version 2 has much fewer figures, but they have been carefully chosen to prioritize the key results. The figures summarize the main results that the author wants the reader to remember from the paper.

A reader could understand the version 2 figures relatively easily, even without reading the legends. The labels are clear, unambiguous, and horizontal. The raw data is shown, plotted on a sensible scale, always starting at zero, making it easier for the reader to judge the data. The colour scheme links the different figures.

The version 2 figures stand alone. The combination of figure and legend would allow someone to understand the figures even if they had not read the rest of the paper, or were returning to it after a long time.

---

[14] Shows raw data and means. Colours are consistent with Version 2: Figures 1 and 2. Labels are horizontal, a good size, and informative (no need to read legend). Legend is short, but includes title and summary sentence. Compare with Version 1: Figure 2. This figure shows a negative result, and so a figure was not necessarily needed. However, the result contrasted with previous work and so was included.

Box 5.6 Take the Introduction Quiz before reading Chapter 6

6

Introductions

The Introduction is one of the most important parts of your paper. When the paper is submitted for publication, the Introduction must convince the referees that the paper is on an important topic and deserves to be published in your chosen journal (Cartoon 6.1). By the time a referee finishes reading the Introduction, they have often formed their opinion as to whether the paper should be accepted or rejected. *You need to sell your story as well as tell your story.*

Cartoon 6.1 *Your Introduction needs to convince the referees that your paper is on an important topic.*

Once published, the Introduction must convince potential readers that the rest of your paper is worth reading. Remember that potential readers are time stressed, and you have to convince them to go on. They are not obliged to finish reading your paper. Why should they read this paper, and not one of the thousands of other papers competing for their attention?

The Introduction can also be one of the most daunting sections to write—so much so, that some writers are paralysed into inactivity. Tough questions that need answers include: What should I include? Do I know enough? How long should it be? How should I structure it? How can I convince a referee or hook a potential reader? Where to begin? Fortunately, we can help you address all of these difficult issues.

Scientific Papers Made Easy. Stuart West and Lindsay Turnbull, Oxford University Press. © Stuart West and Lindsay Turnbull (2023).
DOI: 10.1093/oso/9780192862785.003.0006

In this chapter, we begin by providing a structure that allows anyone to write a good Introduction by focusing on just three key points. By concentrating on these points, your Introduction will provide a clear, structured narrative, that will quickly convince any reader that your paper is important. We then suggest a reasonable target length for an Introduction, provide some bonus tips, and consider how to include citations.

Structure your Introduction

The purpose of an Introduction is to make clear why you did this work, and how it fits into other literature. In other words, to 'introduce' your work to the reader, and convince them that it is an important, useful, and interesting thing to have done. To do this, you need to explain why the general area is interesting or important, but also that there is an outstanding problem or issue that needs to be fixed. Your paper has then fixed that problem.

Funnel your readers

A good Introduction will start very general, and outline the big issue, but will then focus on exactly what the paper is about. Think of the Introduction like a funnel: broad at the top, and then rapidly narrowing (Cartoon 6.2). You need to guide your reader down the funnel, so that they know why your work is needed and roughly what you have done, before they fall into the Methods section.

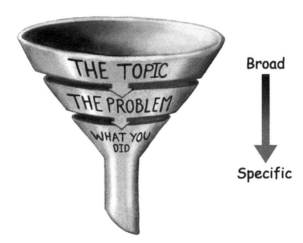

Cartoon 6.2 *The Introduction as a funnel. Your Introduction should be like a funnel, guiding your reader from the broad to the specific.*

Usually, as we go from broad to more specific, an Introduction should cover three things:

1. The area or topic of the paper, and why it is important.
2. A problem in that area—something that isn't known, or a more specific question that hasn't been addressed.
3. What has been done in this paper.

These three points should be linked to each other, to provide a smooth narrative structure. Point 1 is about convincing the reader that the general subject area is important. Point 2 then identifies an outstanding problem within that subject area. Point 3 describes how the paper that you have written will solve the problem identified in point 2.

By following these three points you will naturally move your reader through the funnel, from the general to the specific. Point 1 is very general—the big issue. Point 2 is a bit more specific—a particular problem within that big issue. And point 3 is very specific, saying exactly what you have done in your paper to fix that particular problem.

A simple example: it must be bunnies

So, for example, imagine that your work was examining why trees have been failing to grow in a particular area of the Scottish Highlands. Your hypothesis was that tree seedlings are being eaten by herbivores such as deer and rabbits (Cartoon 6.3). You tested this hypothesis by putting up fences to exclude herbivores for several years, and then you compared the growth of trees inside and outside the fences. In this case, the points in your Introduction would be:

Cartoon 6.3 *Was it the bunnies?*

1. Explain the general issue of why we should care about trees, in the Scottish highlands or more widely.
2. Explain that a possible cause of poor tree growth is herbivores, such as deer and rabbits, but that this idea has never been directly tested.
3. Explain that you directly measured the impact of herbivores on tree growth, by excluding them with fences.

Note that these points link together, and provide a narrative thread that makes sense, and so makes the reader want to carry on reading. In the case of our example, the chain of thought going through the reader's mind will be something like: 'Oh no, why aren't more trees growing?' to: 'I wish someone would test whether it's rabbits' to: 'Brilliant, these awesome scientists are going to test whether it's rabbits!'. And, crucially, by the time they have finished reading this Introduction, the reader will have arrived at the Methods section, knowing why this work is needed, and roughly what has been done.

Answer the *1, 2, 3, Intro* questions

An even simpler way of thinking about the points that need to be covered in an Introduction is to think about the answers to three key questions (Cartoon 6.4):

Cartoon 6.4 *1, 2, 3, Intro!*

1. What is the big question addressed by my work?
2. What still needs to be done in this field (identify a knowledge gap)?
3. How did I address this knowledge gap?

These three questions can be summarized even more briefly as:

1. What is the question?
2. What is the problem? (What needs doing?)
3. What did I do?

In the context of our tree example, the answers are:

1. Why aren't more trees growing in the Scottish Highlands?
2. No one knows whether tree growth is held back by herbivores, like deer and rabbits.
3. We experimentally tested whether excluding herbivores, like deer and rabbits, increased tree growth.

As long as you can answer the three questions, you are most of the way towards writing an Introduction that will be crystal clear to any reader (Box 6.1). We are not saying that this is the *only* way to write an Introduction, but it is one route to a simple and clear Introduction, that will make things easy for your reader.

Box 6.1 1, 2, 3, Intro!

One way to write an Introduction is to break it down into the answers to three questions (Cartoon 6.5). The first question is about identifying the topic or field of research. The second question is about identifying a knowledge gap, or what still needs to be done. The final question is about how you addressed that knowledge gap.

Putting this principle into practice

How do we move from the *1, 2, 3, Intro* structure to an actual Introduction? Easy! In the simplest case, each of our three points/questions will become one paragraph. So, for our tree growth example:

> *The lack of forest growth in Scotland poses a major ecological problem. In both unmanaged grasslands and areas where tree seedlings are actively planted, trees fail to reach maturity. This matters, both for maintaining biological diversity, and for the forestry industry. Many animal species can only survive on mature trees, and the forestry industry relies on a supply of mature trees.*
>
> *A possible explanation for the lack of tree growth is pressure imposed by herbivores. The density of herbivores, such as deer and rabbits, has risen by approximately 300% over the last 40 years in Scotland. These species may be eating tree seedlings, preventing them from*

Cartoon 6.5 *The 1, 2, 3, Intro questions.*

reaching large sizes. There is, however, a lack of data examining the impact of herbivores on tree recruitment.

    We experimentally tested how the presence of herbivores impacts tree growth. We excluded herbivores from experimental plots with the use of fences, and examined the consequences by comparing tree growth within fences to unfenced control plots.

Use a 'however moment'

The example Introduction on tree growth also illustrates how to maintain momentum through this crucial first section of your paper. We start by saying 'this is an important area'. Then, in the second paragraph, we have our 'however moment', when the reader is forced to pause, and say: 'hold on, there is something we don't know'. This 'however moment' is crucial, as it spells out exactly why the paper is needed. You don't actually need to use the word 'however'; you just need to make clear that there is a knowledge gap—something that needs to be found out.

    After the however moment, the reader can start forward again, and although worried that information is missing from the world, they will quickly be reassured that you are going to provide it. By moving smoothly between these three points, you will give the reader enough momentum to get them through to the next section—and hopefully they are now keen to know more about what you did. In short, the Introduction has explained to the reader that there is a problem, and that your paper will fix it.

Although there are other ways to think about how to structure your Introduction, it will be much easier to read when it has a simple, clear narrative. Using the *1, 2, 3, Intro* questions will help you to identify a narrative. When filling in the detail for our tree growth example Introduction, we also applied many of the tips from Chapter 2 (Box 6.2).

Box 6.2 Revisiting the Top Tips

The tips from Chapter 2 can be applied to any section of a paper. The example Introduction on tree growth and herbivory illustrates how several of these tips are especially relevant when writing an Introduction:

- We avoided jargon. Hopefully, you don't need to be an ecologist with a deep knowledge of rabbit herbivory to understand this Introduction.
- We have made one major point per paragraph. The *1, 2, 3, Intro* questions identified three major points, which were each given one paragraph.
- The first sentence in each paragraph provides a summary, or the main point, of that paragraph. Indeed, if all you did was read the first sentence of each paragraph, then you would still get the general picture (try it!).
- We have continually stressed the advantage of planning first, and then gradually fleshing out a 'flow diagram'. The *1, 2, 3, Intro* questions provide the basic structure, which can then be expanded into paragraphs.

Not Introduction

It is useful to step back, and think about what an Introduction isn't for. Remember that the purpose of the Introduction is to introduce the work that you are presenting in this paper. Your job is not to give a deep review of the entire literature, to prove that you know the entire field. Your job is to give a targeted review of the area relevant to the work you are presenting.

For example, if your paper is about foraging in great tits, then you will need to say a bit about foraging, and a bit about foraging in great tits. You don't need to describe everything about great tits, or what we know about foraging in every animal. The *1, 2, 3, Intro* questions force you to focus on the relevant things, and not the unnecessary and potentially distracting things.

Think about different options

In the example on tree growth, we could answer the third question very simply (What did I do?):

We experimentally tested how the presence of herbivores impacts tree growth.

But, depending on the type of paper, it's sometimes more useful to explicitly give your aim, or to state the questions that you are asking in your paper instead. This approach can be especially useful with mathematical modelling papers. For example:

We examined theoretically how plant density influenced the population dynamics of herbivorous insects and their predators. Our aim was to determine whether predators were more likely to die out when plant density was lower. We developed an analytical predator–prey model that allowed the rate at which predators encounter prey to depend upon prey density. We then tested the robustness of our conclusions with an individual-based simulation. This simulation allowed us to examine the consequences of the population being spatially structured into subpopulations, where dispersal was more likely between closer subpopulations.

The extent to which you need to explicitly state your aim or questions will often depend upon the extent to which they are made clear by the description of what you did. Sometimes you might even need to explain one of your results, to help explain why you did something. For example:

We carried out a two-year observational study on the influence of fasting on body weight and fat percentage, comparing individuals who periodically fasted at least four times a week with individuals who never fasted. Our results suggested that while fasting led to a reduction in body weight, it led to a higher percentage of body fat. Consequently, we then carried out a follow-up study, examining whether either exercise rates or nutritional content of food also differed between individuals who did or did not fast.

Unanswered questions

What should a writer do if you don't know the answers to the *1, 2, 3, Intro* questions? It's hard to imagine that anyone doesn't know what they did (question 3), so if you are struggling, it implies that you don't know the answers to questions 1 or 2. To put it another way, you either don't know why the general area is important or you can't identify a clear knowledge gap. If this is the case, then think harder about why you did the work. For example, you must have identified a problem in the literature that needed addressing in order to plan your work. If you can't come up with a satisfactory answer, then perhaps you aren't ready to write up this work yet.

Thinking about the Introduction that you will have to write (and the *1, 2, 3, Intro* questions) can also be a very useful exercise when planning how to conduct your research. Why do you want to work in an area (what is the question?), and what are the outstanding problems in that area (what needs doing?)? If you can't come up with a good Introduction, then maybe it isn't the best project to take forward.

How Long Should an Introduction Be?

Use three to six paragraphs (usually)

The tree growth and herbivory example was deliberately simple. Each of the *1, 2, 3, Intro* questions could be easily answered, and each answer fitted into a single paragraph. Indeed, many Introductions can probably fit into this simple three-paragraph structure, but sometimes things will be more complex, and one or more of the *1, 2, 3, Intro* questions will need an extra paragraph.

For example, you might need two paragraphs to set up the broad research question. The first paragraph could describe a particular phenomenon (such as low forest cover in Scotland), while the second could discuss the possible explanations (herbivores, drought, etc). Alternatively, you might need two paragraphs to describe the problem, or knowledge gap. For example, you might use one paragraph to explain a particular hypothesis, and a second to explain which aspects of that hypothesis have yet to be tested. Finally, you might need two paragraphs to explain what you have done. For example, if the paper has two separate chunks of work, such as some mathematical modelling and some experiments, or two different types of data.

The introduction to this chapter is a good example of when four paragraphs (rather than three) were needed. If you look back to page 95, you will see that our first four paragraphs followed the *1, 2, 3, Intro* structure. The first two paragraphs explained why Introductions are so important (question 1). It was useful to use two paragraphs because we wanted to discuss two reasons why Introductions are important. The third paragraph then explained why Introductions are hard to write (question 2). Finally, the fourth paragraph outlined how we would solve this problem in this chapter (question 3). We know that this Introduction did its job, because it got you interested enough to keep reading and reach here!

Overall, we suggest that around three to six paragraphs should be sufficient for most papers. If your Introduction is at the top of this range, or longer, then take a hard look at it, and question whether you really have used the best structure. Have you kept your focus, or have you gone off target (Cartoon 6.6)? And remember, you can't just cheat by writing enormous paragraphs that go on and on! Of course, there is the usual caveat that we aren't saying you can *never* write more than six paragraphs; we are just saying that usually you shouldn't need to.

Bear in mind that there are always multiple ways to write the same Introduction (and your co-authors might have different ideas). It can be useful to compare and contrast different possibilities, planning them out as a flow diagram before you choose which one to write. And if you want inspiration, then take a look at what other writers have done (Box 6.3).

Box 6.3 How Do Others Write Introductions?

A useful exercise is to analyse papers that you have read recently, that either did or didn't grip you. Look at the Introductions in these papers, and see how they are structured. Do they follow the *1, 2, 3, Intro* style? How long are they? Can you see where edits could be made? It can be really useful to consider how you might restructure or shorten someone else's writing, to help you think about how you would edit your own work.

Cartoon 6.6 *Don't go off target.*

Keep it minimal

The *1, 2, 3, Intro* questions provide a targeted introduction to your work that is brief and to the point. We now want to push this even further, by emphasizing that an Introduction can be improved by making it as brief as possible. The advantages of brevity are to both: (a) make the point and purpose of the paper even clearer, and (b) minimize the chance that we lose the attention of our time-stressed reader.

The key point here is 'as brief as possible' within the context of getting the necessary points over to the reader. You don't want to make the Introduction so short that it doesn't do its job properly. You need to make it just long enough to answer the *1, 2, 3, Intro* questions. By focusing on the key points, it will not only take less time for the reader to read it (which is always an advantage), but it is likely to have a clearer single narrative (because the distractions have been removed).

Perhaps you may feel a certain point has to be made in the paper. If this is the case, ask yourself whether this point needs to be made in the Introduction. Is this point really needed to set up and 'introduce' your work, or could it be moved to a later section, such as the Methods or Discussion? The Introduction benefits hugely from a single clean narrative, while those other sections can better handle side-tracks. For example, caveats about whether you agree with the claims of other researchers, or about how you have carried out your work, can often be moved to later sections.

Bonus Tips

Hopefully, you now feel ready to write an Introduction. You know what to include, what order to put it in, and, crucially, what to leave out. We now provide some bonus tips, to help you take your Introduction to the next level and make it extra-enticing (Cartoon 6.7).

Cartoon 6.7 *An extra-enticing Introduction.*

Open with an interesting sentence

The opening sentence is a great chance to really grab readers. It should be broad, but also interesting. If it's boring, why should the reader proceed any further? Classic examples of bad opening sentences are:

Scientists* have long known that . . . .
Scientists* have long debated that . . . .
There is a long-running debate in science* over . . . .
[*pick your favourite science and insert here]

All these sentences essentially say nothing. If you don't believe they are boring, try saying them without sounding like a robot (good luck!). It should be obvious from the title and abstract what your paper is about, so you don't need to make some vague throwaway statement. If something has long been known, then why are you bothering to say it? Instead, try to orient them more specifically into your subject, tell them something surprising, or set up a tension that needs to be resolved.

Here is an example of how a bland opening statement can be converted into something better.

1. *Tropical forests are known to be very diverse.*

2. *In the lowland forests of Malaysian Borneo, 275 species of Dipterocarp trees dominate the canopy.*

First example—boring! The second example roots the study in a particular place (Malaysia), and focuses attention on a specific group of interest (Dipterocarp trees). Dipterocarps are large rainforest trees, which can be over 80 m in height, and are of major importance for the timber trade. The second example also provides quantitative information that the reader might find surprising, in a way that provokes their interest—for example 'Wow, no way! 275 species from one family. That is crazy! I wonder how they all persist?'.

Now for some more examples of opening sentences. In each case, after you have read the two alternatives, stop and think about which one is better, and why.

1. *Understanding vegetation and the global carbon cycle is of vital importance to the world economy.*
2. *Every year, the combined activity of the world's plants removes some 2,000 gigatonnes of carbon dioxide from the Earth's atmosphere. This service, provided for free, is estimated to be worth somewhere in the region of $10,000,000.*

1. *There is a large literature attempting to explain the damage that parasites cause to their hosts.*
2. Staphylococcus aureus *is a major human pathogen, responsible for 1% of all hospital admissions, leading to an estimated cost of $9.5 billion per year in the United States alone.*

1. *The evolution of multicellular organisms poses an evolutionary problem.*
2. *A blue whale is a cooperative group of about 100 quadrillion cells.*

Explore different possibilities and come up with multiple options for your opening sentence. Judge them on how well you think they: (a) grab the attention of readers, (b) tie in with the rest of your Introduction, and (c) represent what is interesting about your results. A good opening sentence can take many forms.

What will be interesting or novel can also depend hugely upon the knowledge of the audience, which in turn will depend on whether the paper is going to a specialist or generalist journal. Think about your audience. Think about what they may have already heard 100 times already. And think about what they might actually find interesting. A brilliant opening sentence will become bland if it has already been used in the last 100 papers in that area.

Avoid acronyms

Writers often introduce acronyms in the Introduction, to save having to write the same thing multiple times. However, readers will often forget what an acronym stands for, causing them to pause their reading. Perhaps they will eventually remember what it stands for. Or perhaps they will even make the effort to go back and find the place in your paper where you actually defined it. Or perhaps they will just skip on at a cost to their understanding. Acronyms hinder rather than help readers.

Acronyms are only justified in special cases, such as when: (1) the acronym is so common that it is already familiar to every possible reader, like DNA; or (2) the acronym forms part of an accepted name for something, such as a protein or a species of virus. If you really can't be bothered to type out the same thing multiple times, you can always use an acronym, but then search and replace it before you send your paper off for publication. To summarise, *Do Not Acronym (DNA)*.

Consider adding a figure

Don't forget that figures aren't just for your Results section. You can also use figures in your Introduction. Figures can help enormously with clarification, and they can also help to emphasize a key point. We provided an example in Figure 5.3. Other possible uses of a figure in the Introduction include:

1. To explain a complicated concept or hypothesis.
2. To illustrate a specific problem that you are going to solve—either graphically or with photos.
3. To show expected patterns in data under alternative hypotheses.

Most Introductions probably won't need a figure, but it doesn't hurt to step back and think about whether a figure would help. At the same time, remember that adding a figure will take up space, and so the benefit has to be balanced against the cost of the space it will take up.

Add Citations

The Introduction is the first section where you will have to really ponder which citations to add. It's always difficult to know how many citations are needed, and which ones to choose. It's also clear that some authors cite far more papers than others—and while some authors focus solely on original research papers, others prefer to cite reviews. Here we provide suggestions about what to cite and how.

Cite thoughtfully

We suggest four principles to thoughtful citations:

1. *Cite the original primary literature, and not just reviews*. It is important to acknowledge the body of work upon which your research is based. In addition, you need to demonstrate that you know the literature. Knowing the literature is essential to writing a good Introduction, because 'what needs doing' only makes sense in the context of what has already been done. If a referee suspects that you don't know

the literature, they are more likely to question whether you have identified, and made, an important advance.

2. *Cite thoroughly but not excessively.* Your aim is not to cite everything relevant that has ever been written (Cartoon 6.8). Review articles are for comprehensively bringing together the literature. In most original research papers, you only need to include the key, most relevant literature.

3. *Don't overcite.* Don't include too many references for any one point. If the same result has been found 12 times, you don't need to cite all those papers. Instead, just include the first one or two papers that showed the result, a recent review, one or two particularly good examples, or some combination of these (e.g. one or two papers plus a recent review). An exception is when you explicitly need to show that something has been found many times and you are trying to emphasize this point.

4. *Don't cite the same paper repeatedly.* If successive sentences require the same citations, then don't keep repeating them. For example, if one paragraph is describing work from two previous papers, then citing them just at the end of the first sentence in the paragraph could be enough.

As always, the above are just guidelines. The expectations for citations can vary quite a lot across both disciplines and journals. For example, some journals limit the number of citations, so you will have to determine the best combination within that constraint. It's always a good idea to look at how citations have been added to other papers in your discipline, and in the journals where you plan to publish. And look at the instructions for authors on the journal website.

Add citations at the end of writing

We suggest adding citations only once you have finished writing your Introduction, rather than as you go along. Adding citations as you go along can lead to sentences that justify or 'lead into' the references, which can make your writing disjointed and reduce its quality. Instead, focus on writing sentences that flow into each other, to produce a single clear narrative. In summary, write your story first, and then add the citations.

Adding citations earlier in the writing process can also hinder editing. Once a sentence is written, with a citation at the end, it can be hard to completely change it. You can become trapped, and only consider alternatives that fit with the citation. The focus then becomes justifying the citation, rather than the clarity and narrative flow (which are ultimately much more important).

Adding references at the end is a trick to make writing easier. A key assumption behind this trick is that you know the literature well, so will end up citing all or most of the key papers. It's important to know the relevant literature and to try to ensure that you cite people's work fairly. You should not give an unfair or biased representation of the

Cartoon 6.8 *Don't cite excessively.*

literature. If you are worried that you won't remember the key citations and where to put them, then make notes as you go through. The key point is to let your ideas and story guide your writing and overall narrative, not the ideas of individuals.

Put citations at the end of sentences

It can help the reader if you put all your citations at the end of each sentence. Spreading citations through a sentence can make the sentence disjointed and harder to read. Compare:

1. *Organic farming leads to a higher diversity of birds (Benmayor 2020), insects (Leyland 2019), and amphibians (Rowan 2020).*
2. *Organic farming leads to a higher diversity of birds, insects, and amphibians (Leyland 2019; Benmayor 2020; Rowan 2020).*

Summary

Top Tips for an Illuminating Introduction
Structure your Introduction
1. Funnel your readers
2. Answer the *1, 2, 3, Intro* questions
3. Use a 'however moment'
4. Think about different options
Make your Introduction the Right Length
1. Use three to six paragraphs (usually)
2. Keep it minimal
Bonus Tips
1. Open with an interesting sentence
2. DNA: Do Not Acronym
3. Consider adding a figure
Add Citations
1. Cite thoughtfully
2. Add citations at the end of writing
3. Put citations at the end of sentences

Box 6.4 presents a summary example. Take the Discussion quiz before reading Chapter 7 (Box 6.5).

Box 6.4 Summary Example

In this box we give two versions of an Introduction section, for the fruit fly study that we introduced in Box 3.6. The first has problems, which we have annotated. The second Introduction is a better version, based around the tips in this chapter.

Version 1
Introduction

Fruit flies have long been known to be an important agricultural pest.[1] *The increasing rise in pesticide resistance is limiting our ability to reduce the populations of fruit flies and other pests, and hence the damage that they cause. Furthermore, there is an increasing demand for crops which have*

---

[1] If this true, then why need to say? Could the Introduction have instead started with an interesting fact that the reader wouldn't have 'long known'? Also not a great introduction to fruit fly biology.

been treated with no, or lower levels of pesticide.[2] *Consequently, fruit flies are likely to become an even more important agricultural pest in the future.*

Both adult fruit flies (AFF[3]*) and their larvae can interact on fruit. Male flies can use fruit as cues of where to find females. Females can compete to find and/or defend the best sites on which to lay their eggs.*[4] *We know very little about where females choose to lay their eggs. Larvae can compete for food within fruit, with increasing numbers of larvae leading to mortality or reduced size. Higher densities can also help attract or aid parasites.*[5] *For example, parasitoids can be attracted to areas where larvae are at high density. Or, high densities can help spread pathogens.*

Competition is important to all organisms.[6] *Birds and mammals compete for food. Males of many species compete for mates. Insects can compete for areas or 'patches' to lay their eggs. For example, Butterflies require plants for their caterpillars to feed on, and frugivorous insects need fruits for their larvae to feed on. Insects are important agricultural pests, because they eat crops that humans are growing for food.*

Competition can have consequences at many levels of biological organisation.[7] *Population densities can go up and down in response to competition with other species, either directly or indirectly. Individuals can change their behaviour in response to competition, either dispersing to avoid competition or becoming more aggressive in more competitive scenarios. At the community level, competition can shape the community structure and determine the extent to which different species can coexist.*

Fruit flies could also potentially gain benefits from aggregating together.[8] *If there were multiple AFFs*[9] *on a patch, then they could potentially defend the patch better against predators or parasitoids. Also, any consequences of predation or parasitism would be diluted, with individuals gaining from being part of a 'selfish herd'. Larvae could also gain some benefit from sharing a fruit with other larvae, if the movement and feeding of multiple larvae makes it easier to feed in a particular fruit, or if it dilutes other potential threats. In this case, more larvae feeding could lead to higher or quicker development rates and/or bigger larvae.*

---

[2] The background related to pesticide is true, and interesting, but this paper isn't about pesticide, so it is a distraction. The focus on pesticide is especially bad given that this is the first paragraph where the writing should be trying to grab the reader's attention.

[3] Avoid acronyms unless everyone knows them already. They don't really save that much space, and force people to stop and think what they mean—they actively make it harder to read.

[4] This is interesting background on egg-laying behaviour, but we haven't been told why we should be thinking about it. Where did this come from?

[5] The background on influence of density is interesting, but why cover this? Is this essential, or even relevant for what will be done in the study?

[6] The paper is jumping here from more specific (fruit flies) to more general (all organisms). This is taking the narrative in the wrong direction and can be confusing. The reader will be left wondering where they are being taken. Jumping about can be a symptom of the writer trying to review too much information in the Introduction. In addition, how relevant is competition? Does the data in this paper address competition?

[7] There are two major problems with this paragraph. (1) This material raises relatively general issues in the middle of the Introduction, when the writing should be getting more specific (Cartoon 6.1). (2) The work in this paper doesn't specifically address these issues, so it is tangential and could be cut.

[8] This paragraph follows on from some of the things in the second paragraph. But these two parts have been separated by two paragraphs of more general and relatively irrelevant material. Where is the simple narrative flow that leads the reader through? In addition, how much of this material is directly relevant to what is done in the paper?

[9] Can you remember what AFF stood for? Did this acronym make it easier or harder to read the paper?

The costs or benefits of aggregation could vary across fruit fly species.[10] *In cases where flies feed on relatively small fruit, there will be only enough food for a small number of larvae and so aggregation could be relatively costly. In contrast, if flies feed on large fruit, then there can be less competition for resources, which could be outweighed by the benefits of aggregating together. If we could manipulate where females lay eggs then it might be possible to reduce their agricultural impact.*[11] *More generally, the costs and benefits of aggregation could vary hugely across animal taxa, leading to both dispersive and aggregating behaviours.*[12]

Here,[13] *we carried out a number of choice experiments*[14] *in in the fruit fly species Belial lys. We released AFFs*[15] *in a flight cage, and they had a choice of bananas on which to lay their eggs. On some bananas we had caged another mated female, so that our focal experimental females had a choice of bananas at which they would be alone, or where they would be laying eggs alongside another female. Our results are discussed.*[16] *There are a number of potential problems with choice experiments in flight cages, and how these relate to foraging in the real world.*[17]

The first version provides examples of three general problems:

1. It does not follow the funnel—instead it bounces around between general and specific. Consequently, it doesn't lead the reader through a clear narrative. We find ourselves asking: Where are we going, and why?

2. It has a lot of tangential stuff in it, and it is 'padded out' with extra information that is mostly irrelevant. This will try the patience of the reader, but also means that the really key and important things are tucked away and hidden. How can we expect the reader to work out what the key factors are in that stream of consciousness? If any of these things really need to be covered in the paper, then they are probably better left to the Discussion.

3. It does a poor job of really explaining why this paper is needed. Where is the specific problem that we will solve—the 'however moment'?

[10] This is true and interesting, but this paper doesn't test this / take this issue forward, so is irrelevant.

[11] This is a key point regarding why this work has been done. But, it feels hidden amongst irrelevant material. This point is unlikely to jump out of the Introduction. In addition, this point could be better developed: how / why? Where is the 'however moment'?

[12] Irrelevant—this paper doesn't help explain variation in the cost and benefit across species. In addition, the narrative is again going the wrong way from more specific (fruit flies) to more general (all animals).

[13] Words like 'Here' are often pointless to add. Where else would the writing be referring to?

[14] 'Choice experiments' is unclear jargon.

[15] Does this acronym help or hinder the reader?

[16] Of course they are, that is what the Discussion section is for! The Introduction does not need to tell the reader that there will be a Discussion section. You might as well also add: 'We will present our results'.

[17] This point ends the Introduction on a negative note. Plus, this is discussion, only worth entering into after the results of the flight experiments have been given.

Version 2
Introduction

Fruit flies are the greatest pest of agricultural fruit crops.[18] *Adult females lay their eggs in fruit, which the larvae then feed on.*[19] *After a period of several days the fruit fly larvae leave the fruit and pupate, from which they will emerge as adults. In fruit such as melons, papayas, and olives, these flies can cause losses of up to 100% of the crop.*[20]

The damage caused by fruit flies will be greater if females spread out and lay their eggs on many different fruits, rather than aggregating on a small number of fruits[21]. *If we could manipulate where females lay eggs, then it might be possible to reduce their agricultural impact. For example, flies could be lured towards traps, or repelled from certain areas. However, very little is known about how adult females choose where to lay their eggs, and if they have mechanisms to aggregate with or avoid other females.*[22]

We tested whether adult females prefer to aggregate towards other females in the fruit fly species Belial lys.[23] *We carried out a number of choice experiments, where females were released in a flight cage, and they had a choice of bananas on which to lay their eggs. On some bananas we had caged another mated female, so that our focal experimental females had a choice of bananas at which they would be alone, or where they would be laying eggs alongside another female.*[24]

Comparison
Version 2 is less than half the size of version 1. Version 2 is smaller because it focuses on the points that are essential to setting up why we need this paper. It goes through these three points in a simple narrative order, where each subsequent point links to the proceeding point:

1. Fruit flies are a big problem.
2. To address this problem, we need to better understand how fruit flies choose where to lay their eggs.
3. We are going to do some experiments to better understand how fruit flies choose where to lay their eggs.

[18] This version starts with a strong statement, introducing the big issue, that fruit flies are major agricultural pests. An alternative would have to been a fact, like the amount of financial damage caused, but the abstract will be started like that (Box 8.1). It is important to coordinate between the Abstract and the start of the Introduction, so they are not repetitive.

[19] Basic biology, just to make sure all readers are up to speed.

[20] Finishing with a fact to back up the opening sentence. This is a different fact from the one that the abstract will open with (Box 8.1).

[21] This paragraph introduces the more specific problem, that we need to know how females choose where to lay eggs . . .

[22] . . . and that we lack knowledge on this issue. Often the writer can be more specific about the knowledge gap—here, so little is known that it is not worth being more specific.

[23] This paragraph starts by explicitly saying that this paper will do precisely what was said to be needed in the previous paragraph.

[24] This paragraph provides a very simple overview of what will be done in this paper, so the reader can go into the Methods with a general idea of what to expect.

Box 6.5 Take the Discussion Quiz before reading Chapter 7

7

Discussions

Imagine that your reader has made it through the bottleneck of the Methods and Results and is now emerging, with some relief, into the Discussion. The Methods and Results sections are generally quite dense and demanding, as they are filled with technical details, so they are often the hardest parts of the paper to read. The Discussion is where you can return to the bigger picture, and provide a synthesis of your work, leaving your reader with a clear picture of what your paper has achieved (and hopefully wondering who they can recommend it to).

The problem is that providing a synthesis is an open-ended demand, and it isn't always clear what you should include. In this chapter we address key questions such as: What should you discuss? How long should you spend on different issues? How much space should be dedicated to a discussion of other literature or the limitations of your work?

We solve these problems by providing a simple structure, which breaks a Discussion down into three parts:

1. A summary of your results.
2. A point-by-point targeted discussion of the most important findings and implications of your paper.
3. An optional concluding paragraph.

This structure leads to Discussion of an appropriate size, which emphasizes the points that you most want the reader to take away from your paper. We then consider potential pitfalls that need to be avoided when writing a Discussion, and some alternative structures.

Structure your Discussion

A good Discussion will start quite specific with a summary of your results, but then it broadens out, as you put your results into the context of other existing knowledge. Indeed, if an Introduction is like a funnel that gets gradually narrower, then the Discussion is like an upside-down funnel, that gets gradually wider (Cartoon 7.1).

Scientific Papers Made Easy. Stuart West and Lindsay Turnbull, Oxford University Press. © Stuart West and Lindsay Turnbull (2023).
DOI: 10.1093/oso/9780192862785.003.0007

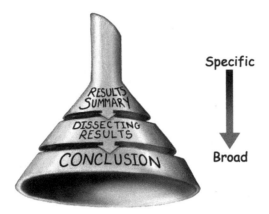

Cartoon 7.1 *The upside-down funnel. Your Discussion should be like an upside-down funnel, taking the reader from your specific results to the broader implications, as they escape from your paper.*

Start with a summary of your results

The first paragraph of the Discussion should briefly summarize the most important findings of your paper. The purpose of this summary is to re-orient the reader. If they have just read the Results section, then they probably feel flooded with data, and this is your chance to make the main points clear to them. What are the key results that you want the reader to really focus on?

But remember that the reader may not have read your Results section. Methods and Results sections can be hard work, so readers often skip from the Introduction to the Discussion, with perhaps a glance at the figures on the way. It's impossible to force every reader to work through your paper methodically! So, by providing a summary of your results at the start of your Discussion, you can make up for a reader's half-hearted engagement and still get your key results over. This can be even more effective if you refer them back to the relevant figures.

Dissect your most important results

After your initial summary, the next few paragraphs should take each of your main results in turn and provide some 'discussion' (Cartoon 7.2). Examples of relevant discussion points include: Do your results agree or disagree with what has been done before? Were the results expected or unexpected? If unexpected, is there some plausible explanation? Are there any problems or limitations that you want to raise?

Use the final paragraph to tie up your paper

The final paragraph can set out the broader agenda. What are the wider implications of your work? Does a whole area of research now need to adapt to your new findings? Do you want to speculate on future work that might be carried out to help resolve your findings with those of others?

Cartoon 7.2 *Dissect your results.*

To summarize, a general outline of a Discussion might look like this:

Paragraph 1: brief and concise summary of your main findings.

Paragraph 2: the most important result dissected and discussed more widely. Was it unexpected, or did it support your hypothesis? Are there problems or limitations with your work that you need highlight?

Paragraphs 3–n: the remaining results treated in the same way.

Last paragraph: Bring all your results back together again. Is there now a new and clear way forward based on your results? What are the obvious 'next steps'?

Start With a Summary of Your Results

We suggest using the opening paragraph of your Discussion to provide a summary of your results. This can vary from an ultra-terse summary of your main results, to something that provides more context and/or implications. We illustrate the possibilities with different versions of the same opening paragraph drawn from a made-up study that had examined how diet influenced the size and body condition of captive hedgehogs.

> *Opening paragraph version A:*
> *We found that when a standard dry animal food was supplemented with worms, this led to hedgehogs: (1) growing to a larger size, with more body fat (Figure 1); (2) having fewer spines on their body (Figure 2); (3) spending less time drinking (Figure 3).*

Version A is an example of a super-terse summary. We have signposted the relevant figures, so that the reader can look back, even if they haven't read the Results section. This version could easily be expanded slightly to include details about the magnitude of different effects. For example, we could have written: '*(2) having 25% fewer spines on their body (Figure 2)*'.

> *Opening paragraph version B:*
> *We examined experimentally how diet influenced the body size and condition of captive hedgehogs. We found that when a standard dry animal food was supplemented with worms,*

this led to hedgehogs: (1) growing to a larger size, with more body fat (Figure 1); (2) having fewer spines on their body (Figure 2); (3) spending less time drinking (Figure 3).

In version B, we have added a sentence at the start of the paragraph to give the reader a gentle reminder of our study design. Providing some context in this way can be very useful. Maybe the reader got lost in the details of the Results and forgot the big picture. Or perhaps they just skipped from the Abstract to the Discussion, and so never had a good idea of what you did. Possibly, they even read the whole paper ages ago, and are now just rereading the Discussion to refresh their memory. Don't forget that there are many ways to read scientific papers.

Opening paragraph version C:
   Ten thousand hedgehogs are currently kept in captivity. We carried out an experimental study to investigate the effects of diet on hedgehog size and condition, to provide clear feeding guidelines for captive animals. Generally, our results matched our expectations, with supplementary worms having clear, positive effects on hedgehog size and body fat (Figure 1). However, there were some unexpected results: for example, supplementary worms led to hedgehogs having fewer spines (Figure 2), and spending less time drinking (Figure 3). Our results underline the need to carry out experimental work under controlled conditions to provide concrete recommendations to zoos that keep wild animals.

In version C, we have added both more context and the implications of our work for different audiences (zoo keepers in this case). This paragraph is now closer to the kind of summary given in the Abstract of a paper, providing information on why the study was carried out, the main results, and the implications of the work. The advantage of this kind of expanded start is that it would orient a reader who hadn't read any other part of your paper.

However, there are disadvantages too, as Option C is longer than Options A or B. Option A is minimal, so the reader can quickly move through it, and there is no 'padding', so it shoves the results clearly into the reader's face. Options B and C are longer, but they are also gentler, providing greater context and a broader summary. At the same time, providing context will repeat earlier parts of the paper, and so could infuriate readers who have carefully read every word of your paper and so don't need it.

The relative advantage of these different approaches depends on several factors. Version C might work well with long and complicated papers, where a reader might have struggled to process the Results section and now needs the equivalent of a breath of fresh air (Cartoon 7.3). Option B might work better for a more general audience, who might only need a brief reminder about why we should care about what you did, while Option A might be best if you have presented a lot of results, but now want to emphasize the most important ones.

The relative advantage of different approaches can also depend on how the rest of your paper is written. If the remainder of your Discussion is short and focused, then it might be more important to explain the overall context, and so Option C could be best. But, if you are going on to discuss each point in detail, then you might want to avoid

Cartoon 7.3 *A tired reader, after struggling through the Results.*

repetition, and so Option A could be better. As ever, it can be useful to plan out the different options and see what you think would work best.

Dissect Your Most Important Results

The next paragraphs in your Discussion will be a dissection of your main results. These dissection paragraphs will clarify what your results mean, and give possible explanations, potential limitations, and links to the exiting literature. You should start with a paragraph that discusses your most important result, followed by a paragraph that discusses your second most important result, and so on.

The writing of these dissection paragraphs can raise tricky questions about which results deserve discussion paragraphs and how much detail is required for each one. To make things easier, we have 10 guiding principles to help you decide what to include and what to leave out:

Order the dissection of your results by their importance

Discuss the most important results first. The reader will then be fresher when they reach them. And given that the reader might give up partway through your discussion, they are more likely to actually read the most important bits.

Allocate space according to importance

The space that you give to an issue will help determine how well the reader takes it in. Don't spend two lines discussing the most important result, and then two paragraphs on some minor result that doesn't really matter.

Stay focused

Your Discussion should be focused on discussion/implications that follow from *your* results. The Discussion is not an excuse for you to say everything that you happen to think about this area of research. You don't have to be as minimal as when writing your Introduction, but don't ramble on.

Follow the earlier sections

The different parts of a paper are not independent—they should flow and lead into each other. Consequently, in the Discussion you should be answering the big questions that you set up in the Introduction. If this isn't the case, then maybe you are not doing your results justice. Or maybe your Introduction needs to be better focused on setting up the problem or problems that you solve!

Acknowledge problems

It can be useful to discuss limitations or problems with your work. No one will know your work as well as you do, and so you are best placed to do this. It is better for you to raise a limitation, and discuss it in a balanced way, than to have a referee raise that limitation, and use it to sink your paper—or to have readers think you are overselling your results. The aim of a scientific paper is to improve knowledge, not to argue that your experiments were perfect.

At the same time, never start downbeat. If your Discussion starts with 'Our results were unexpectedly weak which we attribute to a small sample size', then you have set a negative tone that will doom your paper. If your results are negative, with your explanatory variables having no impact on the variable measured, then just state this, exactly as you would if it was the other way round. It's equally valid. When you go through your results in detail, you can of course comment on any potential problems, such as sample size.

You might want to weave caveats or limitations into different parts of the Discussion, as you go along. Or, you could have a penultimate paragraph which discusses multiple caveats. This can depend upon how many caveats you want to discuss, and how much space you have in the paragraphs.

Consider using subheadings

Subheadings can provide structure that link different paragraphs. In addition, subheadings can break up a long Discussion, making it easier for the reader to work their way through.

Consider ordering by subject

Sometimes it can be better to order your discussion paragraphs by subject rather than by importance. For example, imagine that you had four results, in decreasing order of importance from result 1 to result 4, but that results 1 and 4 were on the same subject or linked in some way, while results 2 and 3 were on a different topic. In this case, you would group discussion paragraphs by subject, and then by importance. Consequently, the order would be:

Paragraph discussing Result 1
Paragraph discussing Result 4
Paragraph discussing Result 2
Paragraph discussing Result 3

You could help the reader by using subheadings to signpost the subject groupings, so the whole Discussion might have the following structure:

Discussion

Paragraph summarizing results

<u>*Subject A*</u>

  Paragraph discussing Result 1
  Paragraph discussing Result 4

<u>*Subject B*</u>

  Paragraph discussing Result 2
  Paragraph discussing Result 3
Concluding paragraph

Keep to a reasonable length

The size required for the Discussion varies between papers. A range of three to five paragraphs is pretty standard, but obviously, if it is a big, complicated paper with more experiments and more data, then you will need more. Papers with some discussion in the Results section, such as some theoretical modelling papers, would require a shorter Discussion section, perhaps as short as one to two paragraphs. The key thing is to think about the reader. What do they really need to have explained and discussed? You want to get the important things over, but as briefly as possible, before the reader gets too bored, and just stops reading (Cartoon 7.4).

Cartoon 7.4 *A bored reader of an overly long Discussion.*

Balance against the length of other sections

Another way of thinking about the length of your Discussion is that your paper should be balanced across the different sections. If you are presenting a small data set then you will need a small Discussion. It will look either mad or megalomaniac if a small two-paragraph Results section is followed by a 10-paragraph Discussion. In general, if you have a Discussion that includes more than five paragraphs, stop and think about whether it really needs to be that long. Perhaps it really does, but an alarm bell should be ringing.

Of course, you can always move some discussion to our useful friend the Supplementary Information. This could include sections that might unbalance the main text or interrupt the narrative, or that might interest a subset of specialist readers only. The Supplementary Information isn't just for figures and tables.

Plan before writing

Of all the sections in your paper, the Discussion is perhaps the most flexible in terms of content. Consequently, it is extra useful to really plan out your Discussion before writing it. If you list the potential discussion points, you can work out the really important ones, that must be included, and separate those from the less important points that can be covered very briefly, or cut.

Planning also gives you a chance to think about the narrative flow of your Discussion. Should you just order your points by importance, or do you also need to group by subject? Or is a discussion of point B needed before you can discuss topic A?

Use a flow diagram or mind map to work out how the different parts of the Discussion are linked, and how best to order them. Subheadings can be used to give structure, or to group different parts of your Discussion together in sensible ways. Even if you don't add subheadings, if you think about what they would be and where you would put them, then it will help you to see how well you have structured the flow of your Discussion. If you haven't structured it well, it will feel like it needs a new subheading every paragraph. *Time spent planning a Discussion before you write it can save a lot of time in the long run.*

In Box 7.1, we show how these guiding principles can be put into practice, to produce the dissecting paragraphs. Of course, at the same time, it is important to keep using the top tips from Chapter 2 (Box 7.2)!

Box 7.1 Example Dissection Paragraphs

In the main text, we have already supplied some different versions of an opening paragraph for the Discussion of a study on hedgehog diet. In this box we provide the paragraphs that then dissect the main results of that study.

Paragraph 2:
We found that when a dry food mix was supplemented with worms, hedgehogs grew 10% larger, and that this increased size was mainly due to increased fat reserves. A potential explanation is that the dry food mix is relatively low in fat (< 1%), whereas worms are high in fat. A previous study by de Witte (2014) did not find that supplementing dry food mix with worms led to an increase in hedgehog body size. The difference between our results could be explained by the fact that de Witte used a different dry food mix, which had been developed for pedigree dog breeds, and contained 10% fat. Our hypothesized role of fat could be directly tested, by experimentally supplementing the dry food mix that we used with fat.

Paragraph 3:
The reduction in spine density of around 10% in hedgehogs fed a worm-supplemented diet was unexpected. However, Klock (1986) reported that wild hedgehogs have fewer spines than captive ones. This difference in spine density between wild and captive hedgehogs is thought to be due to wild hedgehogs mainly eating worms, while captive hedgehogs are usually fed a standard dry mix (Mills 2014; Hood 2017). The dry mix contains high levels of a protein derived from fish, Elasfin 3, which promotes spine growth (Dettmann, 2015). By adding extra worms to the supplemented diet, the diet in our experiment would have been relatively low in Elasfin 3, and hence less stimulating to spine growth. It is currently unclear whether the additional spines of captive hedgehogs are in fact a disadvantage to the animals.

Box 7.2 Revisiting the Top Tips

Several of the tips from Chapter 2 are especially important when writing Discussions:

- As always, use simple, jargon-free language. You want the widest possible audience to understand the implications of your work.
- Make one major point per paragraph.
- The first sentence in each paragraph should provide a summary of, or the main point of, that paragraph. For example, if you only read the first sentence of each paragraph in

Box 7.1, you would get a good idea of the main points. Readers often skim-read, only reading the first sentence of a paragraph—make it easy for such readers.

- Be precise and consistent. If you used a certain word to describe something in the earlier sections of the paper, make sure you use same word in the Discussion. Avoid any potential for confusion.

- Plan, plan, plan! Good planning will lead to a better Discussion, that takes less time to write.

Use the Final Paragraph to Tie up Your Paper

The final paragraph brings things together and ties up the end of your paper. This can be done in a variety of ways. Consider focusing on a general conclusion, an implication for the field, or what needs to be done next. For example, returning to our hedgehog Discussion:

Paragraph 4:
    Taken together, our results indicate that the appearance and health of captive hedgehogs is strongly influenced by diet. Given the large increase in body size when hedgehogs were provided with supplementary worms, we strongly recommend that captive hedgehogs should receive at least two worms per day. As body size is closely correlated with other important traits such as breeding success and lifespan, the additional cost of this live food is, we believe, well worth it (Kraviz 2003). Supplementing with worms is particularly important for the six species of hedgehogs currently held in zoos that are also listed as either endangered or critically endangered. Such animals must be given every chance to thrive in their captive environment.

Note that the final paragraph is for concluding and not summarizing. A summary is already provided by the Abstract. Don't just repeat the points made in earlier paragraphs; rather, *say something new*. Use the conclusion to emphasize a major point or implication of the paper, or to answer the overarching question that was given in the Introduction.

Don't add a final paragraph if it isn't needed

Don't add a final concluding paragraph unless it's needed. Many final paragraphs are filled with statements that don't really say anything new or interesting. For example, writing 'Further research is needed' doesn't add anything, because that will always be the case. If you don't need a final paragraph, you can turn the last of your 'Dissecting Results' paragraphs into the final paragraph. When doing this, it can be good to choose a topic that works at the end, such as potential problems and how to fix them, or future directions. It can also be good to use the last sentence or two to focus on broader or future issues, so that the ending doesn't feel too abrupt.

Finish with a flourish

Try to finish in a lively and memorable way (Cartoon 7.5). The concluding paragraph is the last thing that the reader will read, so try to finish on a positive note. Don't just sort of peter out. A bit like this. Hmmm, I think we've said all the important stuff now … Not sure really … perhaps more work is needed?

Instead, try to finish with a nice snappy soundbite. For example: And now you can see just how easy it is to write a Discussion.

Cartoon 7.5 *Finish with a flourish.*

Pitfalls to Avoid

By the time that you are writing a Discussion, you are often close to finishing the writing of your paper. At this stage, problems can creep in (Cartoon 7.6). Some writers relax and let themselves ramble on. Or they feel the desperate need to take this last chance to say something they have been desperate to say. So, as well as thinking about what to put in a Discussion, you need to think about what to not put in. We make five suggestions:

Don't over-introduce

It can be useful to provide some context at the start of a Discussion. But don't overdo it. If the start of your Discussion section feels like another Introduction then it will frustrate your reader. A reader will want to be heading towards the end of your paper, not starting all over again.

Cartoon 7.6 *Beware of pitfalls!*

Don't present new results

A Discussion is for discussing results, not presenting them. Don't add new data, new results, or new analyses. The reader doesn't want any more technical stuff at this point. It is the Discussion, not the Results section.

Don't overinterpret

You may favour explanation A for your result B. But there are probably other possible explanations, like C or D. It is always better to present your conclusions in as balanced a way as possible. Give your favoured explanation but make clear that there are alternatives.

Don't ramble on

Do not try to make connections between your results and every conceivable previous hypothesis/result in your area of biology. In a similar vein, do not extrapolate too far beyond your particular set-up. It's fairly absurd, for example, to carry out an experiment on tree growth in the jungles of Borneo and then start talking about how to save the Polar Bear. Your results might have relevance for Arctic mammals, but leave it for others to decide. If you feel you have used a 'model system' that has general relevance, then discuss things in a general context. Your Discussion is not a chance for you to give your thoughts on life, the universe, and everything.

Similarly, your Discussion is not for reviewing the literature. Any review of relevant literature that is required to make clear the implications and relevance of your work belongs in the Introduction. Your Discussion is only for discussing your results, and their implications.

Don't go over the top

Do not make over-grandiose statements that are not directly supported by your data. If your paper is really groundbreaking and will change the world, you can afford to be slightly understated about it! Don't *say* that a result or your work is important— use clear and simple writing to *show* that it is important. *Show not say*. If you feel that you have to say something is important, it implies that other people might not agree.

Consider Different Structures

We have already said that there is no single correct way to write a Discussion section. So, step back and think about what works for your paper, and the kind of reader that it is likely to attract. Although we have suggested that it can be useful to order your discussion points by their importance, or to group points by subject, there are sometimes other possibilities that need to be considered.

Consider combining sections

In Chapter 4 we discussed how to combine your Results and Discussion into a single 'Results and Discussion' section. This could be because your paper will work better this way, or it could be a journal requirement.

Consider adding Conclusions

It's usually possible to end your Discussion with a section entitled *Conclusions* or *Concluding remarks*. A conclusion leaves the reader with a summary that they can walk away with. A Conclusions section can be especially useful in a big paper with a long Discussion, or in a paper with a combined Results and Discussion section. In contrast, if it's a short paper, with a short Discussion, then a Conclusion will probably be overkill.

Consider saving an issue for the Discussion to help streamline the Introduction

In Chapter 6, we discussed the advantage of having a minimal Introduction that points out the big question, identifies a specific problem, and then explains how you will solve that problem. A single narrative, that doesn't jump between issues is much easier for a reader to follow.

But sometimes there might be some caveats or complications that it would be disingenuous or misleading not to mention. A useful trick here can be to save these things for the Discussion. Use a paragraph or two in the Discussion for something that needs to be mentioned, but that would disrupt the narrative flow of the Introduction. This trick allows you to have a brief, focused Introduction, that draws potential readers in, while still covering all the essential issues somewhere in the paper. This is an example of blurring the boundaries between sections and is justified because it helps the reader.

Think about the Reader

As when writing any section, the most important thing is to keep thinking about the reader, and what you want them to take away from your paper. What are the main things about your work that you want them to understand? What do you need to explain to them? What research needs to be done next? Stay focused on these key topics.

Summary

Top Tips for a Dynamic Discussion
Structure your Discussion
1. Start with a summary of your results
2. Dissect your most important results
3. User the final paragraph to tie up your paper
Things to Consider
1. You can provide some context for your results summary
2. Order the dissection of your results by their importance
3. Allocate space according to importance
4. Stay focused
5. Follow the earlier sections
6. Acknowledge problems
7. Consider using subheadings
8. Consider ordering by subject
9. Keep to a reasonable length
10. Balance against the length of other sections
11. Plan before writing
12. Don't add a final paragraph if it isn't needed
13. Finish with a flourish
Pitfalls to Avoid
1. Don't over-introduce
2. Don't present new results

3. Don't overinterpret
4. Don't ramble on
5. Don't go over the top

Consider Different Structures

1. Consider combining sections
2. Consider adding Conclusions
3. Consider saving an issue for the Discussion to help streamline the Introduction

Box 7.3 presents a summary example.

Box 7.3 Summary Example

In this box we give two versions of a Discussion section for the fruit fly study that we introduced in Box 3.6. The first version is not good, and we have annotated it to highlight potential problems. The second version is improved, by making use of the tips we have given in this chapter.

Version 1

Discussion

Fruit flies lay their eggs in fruit, which the larvae then feed on. After a period of days the fruit fly larvae leave the fruit and pupate, from which they will emerge as adults. Some species of fruit fly are capable of causing extensive damage to agricultural fruit crops. Ultimately, the damage caused by fruit flies depends upon how and where they lay their eggs. If we could manipulate where females lay eggs then it might be possible to reduce their agricultural impact. For example, if we could induce females to concentrate their eggs on a small number of 'decoy' fruit.[1]

We carried out a number of experiments on the fruit fly species Belial lys, *examining how mated females choose where to lay eggs. In particular, we carried out a number of choice experiments, where females were released in a flight cage, and they had a choice of bananas on which to lay their eggs. On some bananas we had caged another mated female, so that our focal experimental females had a choice of bananas at which they would be alone, or where they would be laying eggs alongside another female.[2]*

We found that females preferentially chose bananas on which we had caged another female.[3] However, not all females chose this option; some chose to lay eggs on an unoccupied banana. Comparing females that laid eggs alongside another female, or alone, we found that females who laid eggs in the presence of another female laid more eggs. In addition, these females stayed on the banana for longer before they dispersed. The rate at which the adult females fed on the bananas, while laying eggs, did not vary depending upon what type of banana they were at.

[1] This paragraph is all Introduction not Discussion. These issues should have been explained when setting up the need for this paper, not here when discussing the results. Could be completely deleted.
[2] This paragraph is Methods not Discussion. It can be useful to give a reminder of what was done, but that can be very brief. This paragraph could be either completely deleted or condensed to one sentence.
[3] At last, some results! It took a while, but we are finally here. The results summary in this paragraph could be shortened, but it provides a reasonable summary. The figures are not cited, which is a pain, as the reader might want a quick look at the data, and so it would be useful to tell them where to look.

Females remained at a banana, and laid more eggs, when there was already a female there. This is likely to be because their larvae do better when there are more larvae. When more larvae are feeding on a fruit, the process of moving and feeding mixes up and softens the fruit, making it easier to extract nutrients. This could lead to better-fed and bigger flies.[4]

Our data cannot explain why females behave this[5] way.[6] Our experiment examined whether females prefer to lay eggs on bananas where there is already another female.[7] We found that they[8] did prefer bananas where other females were already laying their eggs. But we can't explain why we obtained this result. Maybe females are responding to the presence of other females, or maybe they are responding to cues from the larvae of those other females feeding on fruit. Previous work on Drosophila has shown this—that females prefer to feed on fruit where other larvae are already developing. Females might be responding to visual or olfactory cues. More work is needed.[9]

Fruit flies are able to develop on many different fruit.[10] In the agricultural setting, B. lys is an important pest of bananas, mangoes, papaya, and melon. The economic impact is especially large in mango plantations where asynchronous ripening within a plantation leads to large populations building up, and causing substantial impact. This problem can be reduced by spraying pesticide, but this is costly and there are health worries about the pesticide staying in the fruit. Our experiments were on bananas.

Our experiments were not conducted under natural conditions.[11] We examined females making decisions over small spatial scales in cages. In nature, females are likely to be attracted to fruit over much larger distances. We do not know if the same factors apply at these different scales. Further experiments are required.[12]

---

[4] This paragraph is the first dissection of a result. While we need to start dissecting results, is this the best one to start with? It seems more logical to first dissect which fruit they choose, before discussing how long they remained on a fruit. In addition, only one hypothesis is given, as if this must be right. Are there any alternative hypotheses?

[5] The word 'this' is often confusing. What exactly is it referring to? It might be obvious to the writer, but not to the reader.

[6] This sentence is quite a negative start to the paragraph and it continues in a negative tone. There is no need to be so negative—you can't answer everything in one paper. But we have one clear result about behaviour, which then opens new questions, to which there are testable hypotheses.

[7] Why are we getting methods again? Boring repetition.

[8] The word 'they' can be just as confusing as the word 'this'. Better to be more specific and refer to 'females'.

[9] Pointless waste of words. More work is always needed. This adds nothing. Either say nothing or say something more specific.

[10] The information in this paragraph is all very interesting. But the results don't have anything to say about this issue. Could completely delete.

[11] This paragraph has similar problems to one of the previous paragraphs, as it starts very negatively. On one hand, if this work is a first step, which can be built upon, then that is fine—you don't have to do everything in one paper. On other hand, if related experiments were carried out at different spatial scales, then why not put them together into one paper, rather than divide them up?

[12] Another waste of words—further requirements are always required. No paper will finish science! If this paper has highlighted some specific further requirements, then it would be useful to point them out.

To conclude,[13] *we examined how fruit flies choose where to lay eggs. We found that females prefer to lay eggs on bananas where there is already another female laying eggs. In addition, they spent longer on those bananas, and layed more eggs. These results could help us control pest populations.*

Version 2
Discussion

We examined how females choose where to lay eggs, in the fruit fly Belial lys.[14] *We found that, when released into a flight arena, females preferred bananas on which we had caged another female. Specifically, females were more likely to visit, remained longer, and laid more eggs on bananas where we had caged another female (Figures 1 and 2). In contrast, the rate at which females fed on the bananas did not depend on whether there was another female there (Figure 3).*

Our results suggest that females can tell when other females are at a banana.[15] *The presence of other females could be detected via either visual or olfactory cues. Further experiments could test between these possibilities by directly manipulating the available cues. For example, by examining behaviour when they are merely presented with an olfactory flow from occupied and unoccupied bananas. Previous work on* Drosophila *has shown a similar preference for fruit where other larvae are already developing, where the importance of olfactory cues was suggested.*

Females remained at a banana,[16] *and laid more eggs, when there was already a female there.*[17] *One possible explanation for this is that when more larvae are feeding on a fruit, the process of moving and feeding mixes up and softens the fruit, making it easier to extract nutrients. This would lead to each larva doing better at higher larval densities. Alternatively,*[18] *adult females might enjoy benefits from laying eggs alongside another female, such as a reduced chance of predation. These hypotheses could be tested with experiments that directly manipulate densities, and then measure the consequences.*[19]

[13] Is a concluding summary really needed? This paragraph repeats what was said earlier. This paper only includes a relatively simple experiment, so could probably get away with a short Discussion section. And this paragraph finishes relatively vaguely, giving a limp finish.

[14] This version starts with a much briefer summary. One introductory sentence to set the scene and then the results.

[15] This paragraph is a discussion of the most important result.

[16] This paragraph is a discussion of the second most important result.

[17] In this version, the events have been ordered chronologically: discussing where females go to lay eggs, before discussing what they do when they get there (how long they stayed and how eggs they laid). This order feels more logical.

[18] Compared to the equivalent paragraph in version 1, this paragraph discusses multiple competing explanations (hypotheses).

[19] Compared to the equivalent paragraph in version 1, this paragraph says what would be needed next, rather than just 'more work is needed'.

Our results should be seen as a first step in determining how females choose where to lay eggs.[20]
We have examined decisions on a relatively small scale within a flight cage. Whether the same or different cues are used at larger spatial scales could be investigated in a glasshouse, or in the field. Ultimately, if we can understand how females choose where to lay eggs, then we may be able to take advantage of this in pest management; for example,[21] *by creating traps that lure females away from important fruit crops.*[22]

Comparison

Version 2 is approximately half the size of Version 1, but it still manages to cover all the main 'discussion' points. Words have been saved because it is more focused, and it concentrates on the implications of the results in this study. Of course, we aren't claiming that Version 2 is an example of a perfect Discussion. A real Discussion section would probably refer to a bigger existing literature and might have to include more discussion about limitations. But our aim is to show how Discussions can be improved and to make the point that shorter is often better: Version 2 makes the main points much clearer to the reader.

---

[20] This paragraph discusses what isn't known, but in a more positive way. And then it finishes on a positive forward-looking note.

[21] This Discussion finishes on a positive, forward-looking note.

[22] This Discussion is relatively brief, and doesn't need a concluding summary paragraph, as this would just be repeating what was said in the proceeding couple of paragraphs.

8

Abstracts

Abstracts are small but mighty (Cartoon 8.1). Readers use them to quickly form an impression of both the quality and relevance of a paper. Like a good poster or movie trailer, a good Abstract grabs the reader's attention and makes them want to read on, or at least ensures that they leave with the right impression. In contrast, a poor Abstract will turn readers away, no matter how good your science, and can leave a damaging first impression in the minds of reviewers. So, what makes a good Abstract? We outline four key questions that your Abstract needs to answer. By answering these questions, you will write concise, structured Abstracts that are both powerful and informative.

Cartoon 8.1 *Make your Abstract small but mighty.*

Structure your Abstract

Your Abstract will be the first thing people read, and often the last. Will you hold the reader's attention, and motivate them to read on? Only around 1–10% of people who read an Abstract are interested enough to go on and read the whole paper. Even referees—who have to read everything—can form an early opinion about the value of your work from the Abstract. This opinion can be hard to change and might affect how the referee judges the rest of your paper. Consequently, your Abstract can play a fundamental role in both the publication and impact of your paper.

Scientific Papers Made Easy. Stuart West and Lindsay Turnbull, Oxford University Press. © Stuart West and Lindsay Turnbull (2023).
DOI: 10.1093/oso/9780192862785.003.0008

Time spent crafting an Abstract is therefore time well spent. The challenge in writing a really good Abstract is that it should do exactly what it needs to do, *and no more*. This naturally raises many more questions about what exactly a good abstract should do. Should it contain the results, or just the science question (with no 'spoilers')? Does it need to explain your methods, and if so, in how much detail? Should it go into the justification or the implications of your work?

Answer four questions

We suggest that one simple way to structure an Abstract is to provide the answers to four questions:

1. What motivated your work?
2. What did you do?
3. What did you find out?
4. What are the implications?

The answers to these four questions provide a summary of the four main sections of your paper, in the order that they will appear:

1. What motivated your work? (Introduction)
2. What did you do? (Methods)
3. What did you find out? (Results)
4. What are the implications? (Discussion)

An Abstract written in this way provides a simple narrative that will neatly summarize your whole paper. It explains why the paper was needed (just as you would in your Introduction), what you did (Methods), what you found out (Results), and the implications of your findings (Discussion). If you do this as concisely as possible, then your Abstract can't fail to grab the attention of potential readers.

Write your Abstract last, but not least

Our suggested structure emphasizes that your Abstract is really your paper in miniature (Cartoon 8.2). It provides a condensed summary of your work; indeed, some journals even prefer to use the term Summary rather than Abstract. It therefore follows that writing a really good abstract is hard: it needs to do all the jobs of the other sections, but in an incredibly limited space.

Furthermore, it isn't enough to just answer the four questions robotically and with no real effort. Instead, your Abstract needs to be interesting, so that the reader will want to delve into the rest of the paper. As ever, we have to remember that this reader has a million other things that they could be doing, and it is your job to make them want to read

Cartoon 8.2 *An Abstract is a paper in miniature.*

your brilliant science (Box 1.1). And if they do not read on, you at least want them to gain an accurate sense of what you did and what you found out. It's surprisingly common for researchers to cite papers without having read more than the abstract (although please don't accuse your supervisor or colleagues of having done this).

It's often best to save writing the Abstract until the end, because if you have all the other sections sorted, then you know exactly what you need to summarize. However, this strategy of writing the abstract last can make it tempting to do a rushed job. By this stage you may be desperate to just finish and submit your paper. But remember, it really is a case of *last but not least*, and to keep your Abstract *small but mighty*.

A simple example

To give a specific example, let's return to the hypothetical tree growth and herbivory paper from Chapter 6. If we wanted to write an Abstract for this paper, then possible answers to the four questions are:

1. *The lack of tree regrowth in the Scottish Highlands poses a problem both for maintaining biodiversity and for the forest industry.*

2. *We experimentally excluded two potentially important herbivores, rabbits and deer, to test whether they impede tree growth.*

3. *We found that, over a five-year period, the exclusion of these herbivores led to a 20-fold increase in the number of trees establishing to 1 m tall.*

4. *Our results suggest that the widespread use of fences to exclude herbivores could solve the problems surrounding the lack of tree regrowth in the Scottish Highlands.*

These four sentences roughly correspond to the four main sections of the paper:

1. Introduction.
2. Methods.
3. Results.
4. Discussion.

This example also illustrates how Abstracts often start general (the big problem), then get specific (the details of what you did and found out), before getting general again at the end (the implications). This is precisely because an Abstract follows the structure of a paper: Introduction → Methods → Results → Discussion (remember the funnels in Cartoons 6.2 and 7.1).

A more complicated example

The example above was relatively simple, and we were able to answer each of the four questions with a single sentence. But in longer or more complicated papers, you might need more than one sentence to answer one or more of the questions. For example, perhaps you have two main findings to report and these can be reported as separate points. Or perhaps you need multiple points to set up why your work was needed, or maybe your results have both specific and general implications. Consequently, you might need more than four sentences in total. For example:

1. *The bacteria* Parvus bona *is a widespread human pathogen causing 30 million deaths per year. The growth and virulence of this pathogen rely upon the production of molecules that the bacteria secretes to suppress the human immune system.*
2. *We examined the genetic basis of this immune-suppression molecule both in laboratory cultures and in clinical infections.*
3. *We found that the production of immune-suppressing molecules was regulated in response to the level of a small signalling peptide also released by the bacteria. In particular, high bacterial densities led to a build-up of the small peptide, which triggered a 100-fold increase in the production and release of immune-suppression molecules.*
4. *The development of mechanisms to disrupt this signalling system offers a novel intervention strategy to treat P. bona.*

In this example, we have devoted two sentences to the Introduction (Question 1), one to the Methods (Question 2), two to the Results (Question 3), and one to the Discussion/Implications (Question 4). The number of sentences required for each part of an Abstract can vary greatly, depending upon the paper. When you write your Abstract you will need to think about how much space each question requires.

Eight Guiding Principles to Mighty Abstracts

Hopefully you feel ready to plan the structure of your Abstract. But how should this structure be fleshed out to make your Abstract as mighty as possible? We suggest eight guiding principles.

Focus on the important

Your aim is to provide a summary that will interest potential readers and draw them in. You don't have to explain everything you did or everything that you found out, or even every possible implication. You can just focus on the major points. If someone could only take away a couple of things from your paper, what would you want those things to be? Readers who really want all the details can read on to find out more.

Keep it short and punchy

A focused punchy Abstract will draw in more readers than a long, unwieldy one (Cartoon 8.3). The addition of too much information tends to reduce rather than increase clarity, by making it harder for the reader to get the big picture.

Cartoon 8.3 *Make your Abstract short and punchy.*

Link your answers

While we have emphasized that an Abstract provides the answers to four questions, these questions are not completely isolated from each other. The answer to each question feeds into and sets up the next. These links raise the same issue of narrative flow that we emphasized in Chapters 2 and 6, but it can be even harder to achieve when condensed into a single paragraph. An Abstract tells the entire story of your paper. *Make it a clear story.*

Consider combining answers

While we have suggested that an Abstract should provide the answers to four crucial questions, this can be done in different ways. For example, it might be useful to combine your methods and results in a single sentence:

> *We compiled data from the literature on 1,212 bird species and found that species which lived in tropical climates laid more eggs per year.*

Or it might be a good idea to combine your results with their implications:

> *We found that all the woodlice were under the leaf litter, suggesting that they prefer dark damp environments.*

In both these cases, combining the answers to two of the questions into a single sentence contributes to a shorter and more punchy Abstract.

The start of this chapter is another example of how it can be useful to combine sentences. If you look back, you will see that our first paragraph was an Abstract that followed the four-question structure. The first five sentences explained the motivation for writing a good Abstract (question 1). It was useful to do this over five sentences, rather than one, because we wanted to convince you of the importance of reading this chapter. We then used only two sentences to answer the other three questions, by: briefly outlining our suggested method (the four-question structure); stating our results (that this would lead to concise and structured Abstracts) and their implications (people would be more likely to read your paper and get the right message from it). You carried on reading, so presumably our strategy worked!

Avoid details, unless they help

How much detail should you give in an Abstract? Sample sizes? Study locations? Species names? We suggest avoiding details, unless they will help to draw in readers. Remember that the job of an Abstract is not to explain all the details of your paper. You have plenty of room for details in later sections. In the Abstract, you just need to give a brief overview, to convince the reader to go further. Details may even hinder at this stage, by distracting the reader from the big picture.

The exception to this rule is when the details are real selling points of your paper. For example, maybe the sample size was exceptional—perhaps you combined data from 237 previous studies or compared 4,707 different species of birds; or maybe the bacteria species you worked on is of medical importance; or perhaps you carried out your work in a country where this study was not previously possible. In all these cases, it's probably a good idea to include the relevant details as they will clearly help to attract readers.

Avoid jargon

The Abstract is the section most likely to be read by non-specialists, so it's especially important to avoid jargon, which will put them off. In later sections you at least have the space to explain your jargon, which makes it easier to understand. Do not under any circumstances introduce a new acronym into an Abstract!

Avoid direct repetition

As tempting as it might be, it is a bad idea to just cut and paste sentences from other sections into your Abstract. It's true that you are summarizing all the other sections, but it's boring for the reader to encounter the same sentence or sentences.

The temptation to repeat is especially strong with the opening sentences of the Abstract and the Introduction. If you have found a good way to 'open' your paper, you might wonder why you can't use the same sentence for both sections. The answer is that not only would this be boring for the reader, but it is also a wasted opportunity—you could instead be making a different point, or making the same point in a different way. For example, if you have two different contenders for an opening sentence, then you can use one for the Abstract and one for the Introduction. This can allow you to have a single clear narrative in the Introduction, but also squeeze in another angle in the Abstract.

Never start downbeat

Don't set a negative tone that will put readers off. While it is important to be open about limitations or problems with your work, this is usually better done in the Discussion than the Abstract (Chapter 7).

Grab Your Reader's Attention

The first sentence or two of an Abstract are the first thing that a reader is likely to read. So, it's worth spending some time to make sure that the opener grabs the attention of your reader and is crystal clear (Cartoon 8.4). Explain your motivation and say something interesting.

Cartoon 8.4 *Grab the reader's attention!*

Explain your motivation

The first part of your Abstract (question 1) is analogous to your Introduction section, explaining what motivated your work. In Chapter 6, we suggested starting an Introduction by explaining two things:

(1) The area or topic of the paper (what is the question?).
(2) A specific problem in that area (what is the problem?).

It can be useful to start an Abstract in the same way. For example:

> *Many pathogenic bacteria use between-cell signalling to coordinate attacks on the immune systems of their hosts. However, we do not know the genes that control this signalling.*
>   *Evolutionary theory predicts that individuals should preferentially cooperate with closer relatives. However, numerous studies on insects and mammals have failed to find support for this prediction.*

In both these examples, we first identify an area, before highlighting a more specific problem within that area (the 'however moment'). This creates the all-important narrative flow.

Start with an interesting sentence

In Chapter 6, we discussed the importance of the opening sentence in an Introduction. Everything we said there also applies to the opening sentence of an Abstract, but even more so. Your opening sentence is an amazing opportunity to pull in potential readers. Use it!
   To give an extreme example, here are two possible opening sentences for a paper on how wasps control the sex of their offspring:

1. *Evolutionary theory predicts that, when mating occurs between the offspring of a small number of females, those females should bias their offspring sex ratio towards females.*
2. *The sex ratios produced by parasitoid wasps in the genus Melittobia are scandalous.*

Which option do you prefer and why? Option 1 provides the theory. It is a solid introduction and a perfectly valid way to start, but it's also a little dull. In contrast, Option 2 is perhaps less clear, but it certainly grabs the attention. Even if you knew nothing about sex ratios, you might still want to read on and find out why they are scandalous. Here are two more examples:

1. *Bacteriocins are small peptides that possess antimicrobial activity against other bacteria.*
2. *Some bacteria wage 'chemical warfare' by using small peptides called bacteriocins to eliminate competitors.*

1. *Explaining cooperation is one of the greatest challenges for evolutionary biology.*

2. *Animals living in harsh environments, where temperatures are hot and rainfall is unpredictable, are more likely to breed in cooperative groups.*

In each of these examples, ask yourself whether your preferred version 1 or 2, and why. If you are writing a paper yourself, you can also try to produce multiple versions, to see which you (and your colleagues) prefer.

Our advice is to try and make the first sentence interesting and easy to understand, but also to avoid making it too predictable or boring. By predictable, we mean don't use a sentence or a variant of a sentence that has already been used to start a hundred other papers in your area. But be careful not to overdo it.

A good test is to read your first sentence or two out loud (Cartoon 2.1). Does it sound interesting, or do you sound like you're trying too hard? We started this chapter with 'Abstracts are small but mighty' to try and convince you to read on. Perhaps you found that a little cheesy—but it's all a matter of taste (Cartoon 8.5).

Cartoon 8.5 *It's all a matter of taste. Try multiple options, and see what you prefer.*

To give an example that is certainly original, but perhaps not what we had in mind, here is an opening sentence from a paper in an interdisciplinary journal: *Proceedings of the National Academy of Sciences*, where the articles are supposed to be aimed at non-specialists:

Among the several central meanings of Darwinism, his version of Lyellian uniformitarianism—the extrapolationist commitment to viewing causes of small-scale, observable change in modern populations as the complete source, by smooth extension through geological time, of all magnitudes and sequences in evolution—has most contributed to the causal hegemony of microevolution and the assumption that paleontology can document the contingent history of life but cannot act as a domain of novel evolutionary theory.[1]

Now ask yourself whether you would you want to go on and read this paper? The opening sentence certainly isn't predictable, but even an evolutionary biologist would be left confused, let alone a non-specialist. If the first sentence of the Abstract is hard to follow, then readers will assume that the rest of the paper will be the same, and just give up reading it.

In summary: the first sentence of an Abstract can have a huge influence on the number of people that read a paper. Try to make it interesting, but also clear.

Different Styles of Abstract

So far, our suggestions have been aimed at those writing a relatively generic Abstract. Most journals require a single paragraph, which can be formed by joining the answers to the four questions together. Some journals, however, have different styles and/or word limits.

Check the journal style

In some journals the Abstract is divided into bullet points or a numbered list. If you have to write an Abstract divided into several points, then it's usually best to make sure each point only addresses one of the four questions: (1) motivations; (2) methods; (3) findings, or (4) implications.

So, for example, if you are allowed six points, then you might go for something like this:

1. Motivation
2. What I did
3. Finding 1
4. Finding 2
5. Specific implications
6. More general implications

Or:

1. Motivation
2. More about motivation

---

[1] Gould, S. J. (1994) Tempo and mode in the macroevolutionary reconstruction of Darwinism. *Proceedings of the National Academy of Sciences* 91: 6764–6771.

3. What I did

4. Finding 1

5. Finding 2

6. Implications

It's important to remember that you don't have to use every available point. Perhaps you can summarize your paper with fewer points, which would be great!

If you are allowed multiple sentences per point, then you might want to group them according to the four questions. For example:

1. Motivation

2. What I did

3. Finding 1. Finding 2

4. Specific implications. More general implications

Some journals state exactly what should be included in each point. Other journals break an Abstract into subsections. For example, *Evolutionary Ecology Research* breaks the Abstract into: Background, Hypothesis, Organisms, Methods, and Results. Many of these structures align very closely with our suggested four questions.

Some journals restrict you to 150 words, others to 350, while some journals impose no word limit at all. In all cases, we suggest trying to keep your Abstract as concise as possible. Your job is to draw in readers, which is helped by brevity and not using up the word limit.

Given this variation across journals, it's also only worth writing the Abstract once you have decided which journal to send your paper to. Looking at how different journals constrain Abstracts can be useful for thinking about what to put into an Abstract and, crucially, what to leave out. All journals are trying to achieve the same outcome, which is to take the main point from each section of your paper and run them together into a concise summary.

As ever, the best strategy is to think of your reader, and try to work out how you can give them what they need, within the constraints of the journal.

Think about a Graphical Abstract

Some journals request a 'Graphical Abstract'—a single-panel image to summarize the main take-home message of the paper. Even if the journal doesn't request one, this can still be a very useful thing to do: you could include it in the Discussion, or the Supplementary Information, or on Twitter when announcing the paper's publication, or even in presentations when you explain your work in person.

A good Graphical Abstract can be an awesome tool for attracting readers, via routes such as the journal's website or social media. It's easy to feel that, by the time your paper is published, you have had enough of it already and don't want to invest any more effort. But when you consider how long it takes to produce a piece of published science, the extra effort required to publicize it is worth it, as it can really increase the number of people that read your work (which should be your ultimate goal).

To make a good Graphical Abstract, you still need to follow the guidelines from this chapter, as well as those in Chapter 5 on figures. But, because you are now summarizing your paper with a single graphic, you need to take those guidelines to the extreme. Although the details can vary across papers, we suggest:

- Pick a single message that you want the readers to take from the figure.
- Keep the figure as simple and clear as possible.
- Cut anything that doesn't help with the take-home message.
- You don't need to set up your motivation—it is only about what you found out.
- Use simple labels.
- Minimize text.
- Avoid adding things that distract from or clutter the main image.
- Think broadly or laterally—what can you do to make it fun, or liven it up for the reader?
- Test-drive your Graphical Abstract on friends who aren't experts in the area.

Some journals also allow a Video Abstract—a very short video to summarize the paper. And again, even if they don't, this could still be a useful thing to do, for social media, or a website. The same considerations apply as for Graphical Abstracts, except that there is usually enough space in a video to allow a structure more like a written Abstract, giving you the chance to state your motivation, what you did, and what you found out. There are even YouTube channels aimed at providing summaries of papers.[2]

With both Graphical and Video Abstracts, it is crucial to think about your reader or viewer. Remember that tired, stressed, bored reader? Well, this is a great chance to make it more fun and interesting for them—so don't just say or present things they will already have seen a hundred times (Cartoon 8.6).

Start working on your cover letter

Both the Abstract and the cover letter, which will explain your paper to the journal, involve summarizing a paper, and so it can be efficient to write them at the same time. Consequently, while you are writing your Abstract is also a good time to start planning your cover letter, which we conveniently cover in Chapter 10.

---

[2] For example 'Two Minute Papers' (https://www.youtube.com/c/KárolyZsolnai/featured).

Cartoon 8.6 *An intrigued reader.*

Summary

Top Tips for an Awesome Abstract

Structure your Abstract

1. An Abstract is your paper in miniature. Small but mighty
2. Answer four questions: (i) What motivated your work? (ii) What did you do? (iii) What did you find out? (iv) What are the implications?
3. Write your Abstract last, but not least

Eight Guiding Principles

1. Focus on the important
2. Keep it short and punchy
3. Link your answers
4. Consider combining answers
5. Avoid details, unless they help
6. Avoid jargon
7. Avoid direct repetition
8. Never start downbeat

Grab your Reader's Attention

1. Explain your motivation
2. Start with an interesting sentence

Different Styles of Abstract

1. Check the journal style
2. Think about a Graphical Abstract
3. Start working on your cover letter

Box 8.1 presents a summary example.

Box 8.1 Summary Example

In this box we give two versions of an Abstract for the hypothetical fruit fly foraging research that we introduced in Box 3.6. The first version has problems, and we have annotated it to highlight some of them. The second version is based around our suggested four-question structure.

Version 1

We carried out a study on the foraging behavior of Belial lys,[3] *in an attempt to understand how females choose where to oviposit[4].[5] We could manipulate choice of oviposition site if we knew what factors caused them to choose certain sites and reject others.[6] Fruit flies cause considerable damage to a range of agricultural crops. Previous studies on fruit fly foraging have been restricted to simple choice tests between different fruit.[7] We tested how females chose oviposition sites (in a flight cage),[8] depending upon whether they could oviposit alone, or alongside another female (on a banana).[9] We found that females did adjust their choice[10] of oviposition site in response to the presence of another female (on a banana). In particular, females were more likely to oviposit (on a banana) if it already had a female caged there.[11] Our results supported the hypothesis that females prefer to oviposit in the presence of other females.[12] We do not know[13] if our results are due to females making use of olfactory or visual cues of the presence of other females (on a banana), or the underlying selective factors which have favoured this behavior.[14] We also found[15] that the rate at which females fed on bananas did not vary depending upon the presence of another female (at that banana).[16]*

[3] What is *Belial lys*?

[4] 'Oviposit' is jargon which readers might not know, in the first sentence! *Avoid jargon.*

[5] This first sentence is methods, before we have been told why we should care about this.

[6] It is good to explain how we can use the data, but we still don't know why we would want to manipulate choice. In addition, the Abstract never really returns to how the results of this paper could help us manipulate choice (i.e. it sets up a question as important, but never answers it).

[7] So what? How have choice tests limited what we know?

[8] Avoid brackets if possible, especially in the middle of a sentence.

[9] Avoid brackets if possible.

[10] Just tells us there was an effect, doesn't tell us in what direction. *Use simple language, not statspeak.*

[11] This sentence gives the result before describing the method. You need to explain what you did before what you found.

[12] This sentence gives the hypothesis after the sentence in which the result was given. It is also repetitive.

[13] This is quite a negative way to present the results. *Never start downbeat.*

[14] This sentence has too much in, and is unclear.

[15] This sentence is returning to describe results, after a sentence on implications. This jumping about is likely to confuse the reader.

[16] At least Version 1 did include results. Some real abstracts are so uninformative that they don't tell the reader about the results!

Version 2

Fruit flies cause $4 billion worth of damage annually to agricultural crops.[17] *If we could manipulate where females choose to lay eggs, then this damage could be reduced. In natural conditions, female fruit lies tend to aggregate at fruit, but we have no idea how the presence of other females influences their choice of where to lay eggs.*[18] *We examined how the presence of other females influenced this behavior, in the fruit fly Belial lys.*[19] *We found that, when released into a flight cage, females preferred to lay eggs on bananas where we had caged another female. This suggests the potential of developing visual or scent-based traps, which could be used to lure females away from fruit crops.*

Comparison

Version 2 is about half the size of Version 1, but does as a far better job of summarizing the paper, in a way that allows the reader to grasp the main points. Version 1 might seem overly silly, but it really isn't. Jumbled Abstracts are common, especially in journals that allow longer ones.

---

[17] This opening sentence gives a fact that might interest the reader. *Grab your reader's attention.*

[18] This version has used three sentences to set up the motivation. Sometimes setting up the motivation in a nice gentle way takes time. But this is time well spent, because if the reader doesn't understand the motivation, they won't care about what you did and what you found out. In this case, it doesn't matter, because the following bits are nice and brief (one sentence each).

[19] It was worth using the Latin name here, for this well-studied pest, to draw the attention of other researchers who work on this species. The common name is given first to avoid scaring away researchers who do not know the Latin name of this species.

9

Titles

Once your Abstract is written, it's time to start thinking about the Title of your paper. You already know exactly what your paper contains and have written an Abstract to summarize the most important points. The Title simply needs to do the paper justice.

So, why is the Title important? As ever, it's useful to put yourself in the place of a potential reader. They are probably scanning lots of papers, trying to decide which ones to read and which ones to instantly forget (Cartoon 9.1). If a Title stands out, then they might take the next step and actually read the Abstract (and maybe even the rest of your paper!). But if your Title doesn't interest them, they will just move on to the next paper, and likely never look back—there are too many papers for them to read anyway.

Cartoon 9.1 *A potential reader, scanning papers.*

In this chapter, we provide a few helpful guidelines, to help you write a Title that is short, snappy, and gripping. A Title that will leave the reader thinking: 'that sounds interesting, I want to know more'.

Scientific Papers Made Easy. Stuart West and Lindsay Turnbull, Oxford University Press. © Stuart West and Lindsay Turnbull (2023). DOI: 10.1093/oso/9780192862785.003.0009

Try Different Options

Title styles vary enormously across different journals and different fields of research, so there is no single winning formula. In addition, the Title of any single paper could take many forms. Consequently, we suggest that you explore different possibilities and experiment. The Title is a great place for trying out different options, because Titles are short, so it doesn't take too much effort.

Produce a list of around five candidate Titles to share with your co-authors (including a couple that perhaps you're not so keen on). Then ask them which one they think works best and why. Discus the options and choose the winner. And if you're still having trouble deciding, here are a few things to try:

1. Read your different options out loud.
2. Rule out any that are so boring that you end up sounding like a robot when you say them.
3. Leave your shortlist for a day or two and then return to it.

Remember that your Title is the first hurdle that a reader must cross to get into your paper, so there can be a large payoff to putting in a little extra effort.

Guiding Principles

Titles are relatively short, often with a maximum wordcount, so you won't be able to fit in everything. For example, you won't be able to mention the question, methods, results, and implications. Instead, you need to be very choosy about what you include—it is a Title, not an Abstract.

In this section we provide eight guiding principles for producing a Title in a variety of styles, that focus on different aspects of your paper. By following these guidelines, and experimenting, you should be able to produce a selection of possible Titles for your paper, from which you can choose the best.

Convey your message, not your methods

Your Title needs to convey the message of your paper: what you found out. It is usually better to focus on your results or the take-home message, rather than your methods. You want to convey the excitement of your discovery, not how much work you did to get there. Compare:

1. *A comprehensive space telescope survey was carried out searching for signs of life on planets 10–50 light years from Earth*

2. *Signs of life were detected on a planet 28 light years from Earth*

If the purpose of your work is to develop a new method, the title could focus on that method, but then you should explain why this new method is useful. Compare:

1. *A neural network method to predict protein structure*
2. *Neural networks double the accuracy of protein structure predictions*

Be specific

Don't be vague. It's essential to leave the reader with a clear impression of what you found out. Just like in your Results section, state the direction of any effects, not just the bare fact that you found something out (Chapter 4). Maybe also give the magnitude of your result. Compare:

1. *Human density is linked to changes in the population size of lions.*
2. *Lion population sizes are declining faster in areas with higher human density.*

Make the message clear

Make your central message as clear as possible. This will often involve putting the message at the start of the Title, and not in the middle or at the end. Compare:

1. *Long-term monitoring of natural populations has found that mosquito populations in Panama have been reduced to the point of elimination*
2. *Biocontrol strategies effectively eliminate mosquitoes in Panama*

Don't try to do too much

Sometimes there will be multiple interesting messages in your paper. It might be possible to capture more than one of these in your Title. But, if not, you will face a tough decision, about which message is most important. Compare:

1. *Plant diversity was greater in habitats with a higher annual rainfall and when herbivore density was higher*
2. *Plant diversity was greater in habitats with a higher annual rainfall*
3. *Plant diversity was greater in habitats with a higher herbivore density*

Again, try different options. Perhaps different options will attract different readers. If so, what kind of readers do you most want to attract? *Choose your readers!*

Be concise

Remember that the job of a Title isn't to convey everything that you did and/or found out. A Title just needs to get over your main message in a way that will attract readers. So, make your Title as short as possible, while keeping the message clear. Compare:

1. *A study showing that spraying with pesticide in orchards makes the flowers on five species of apple trees less likely to be pollinated by foraging honey bees*
2. *Spraying with pesticide makes the flowers on apple trees less likely to be pollinated by honey bees*
3. *Pesticides reduce the pollination of apple flowers by honey bees*
4. *Pesticides reduce flower pollination by bees*
5. *Pesticides reduce the pollination of apple trees*

In this series of titles, we've gradually removed information. There is usually a trade-off between content and readability, so versions 3 to 5 are much easier to digest. They do provide less information, but their simplicity and brevity make them easier to read.

Sometimes extra words can provide further information, which can help the reader. But they can also make the Title more cumbersome and harder to read. At the very least be ruthless and omit needless words. Remove phrases that are so obvious that they don't need saying, such as 'A study of ...' or 'Research into ...'. If it wasn't a study or research, you wouldn't be writing it up. As a rule of thumb, five to 15 words is a good target length.

The advantage of being concise has even been quantified—papers with shorter titles receive more citations.[1] Remember, it is a title, not an abstract—you don't need to explain everything, just capture the main idea.

Keep it simple

While it's important to be precise, you need to make sure your Title can be understood by as many readers as possible. Use simple English (Chapter 2). Avoid jargon and technical terms. Remember that technical terms with which you are very familiar can be impenetrable jargon to others. If a reader struggles with your Title, they are unlikely to go further into your paper. Compare:

1. *Topological constraints drive transitions to cooperation*
2. *Cooperation is favoured in small groups*

As a scientist, you probably worry more than most people about being accurate and precise. However, do not be frightened to simplify your findings (Cartoon 9.2). A key skill in communicating science is to know when to simplify the message. Do not get hung up on the details or exceptions—that's what the rest of your paper is for. Compare:

---

[1] Latchford, A. *et al.* (2015) The advantage of short paper titles. *Royal Society Open Science* 2: 150266.

Cartoon 9.2 *Don't be frightened to simplify your findings.*

1. *The breeding success of red-crowned cranes is influenced by the interaction between nest site provision and carnivore density*
2. *Providing nesting sites increases the breeding success of red-crowned cranes*

You must be careful not to actually lie in your Title. In the case above, the simplification doesn't lead to the Title being wrong, but the real message is clearly more nuanced. In a Title, this is acceptable.

Avoid acronyms and abbreviations

We have already outlined why acronyms and abbreviations can hinder the reader (Chapter 6). The problem is far worse in a Title, where the reader has less explanation and context. Compare:

1. *HGT has facilitated the spread of antibiotic resistance in hospital infections*
2. *Horizontal gene transfer has facilitated the spread of antibiotic resistance in hospital infections*

If the reader didn't already know what HGT is, then version 1 would make them think the paper wasn't on a topic of interest to them. If you use an acronym in a paper, you are effectively limiting your readers to those who are very familiar with that acronym.

There are no hard and fast rules

All the above are just guidelines; they are not hard and fast rules that can't be broken. For example, perhaps the Methods were a really key part of your paper. Possibly they are novel and exciting, allowing you to do something that has never been done before, and so will draw in readers. In this case, you should showcase them in your Title. Compare:

1. *Fungi move phosphorous around their network towards plants that obtain less phosphorous from the soil*

2. *Quantum dot tracking shows that fungi move phosphorous towards plants with low phosphorous supplies*

In the above example, the quantum dot tracking was a new method, worth mentioning. But we still included the message as well. To just focus on methods would have been a bit boring:

Using quantum-dot tracking in fungi

As a test, ask yourself whether the major aim of your paper is to communicate your method, or your findings. If you used standard methods, then they probably should not be in your Title.

Fancy Flourishes

Our above guidelines will hopefully help you to produce a few different possible Titles for your paper. In many cases, a short, simple description of the main message (result) will be the best option. However, there are some alternatives that can be useful in special cases. In all of these cases, it is important not to oversell your paper.

Consider using a question

Sometimes it can work to have a direct question as a Title. This approach emphasizes the need for answers, rather than focusing on what you found out. If you pose a really interesting question that grabs the reader's attention, then they will want to read on—in the hope that you will provide the answer. For example:

What limits the size of insects?
How do communication systems emerge?
Why have some birds lost the ability to fly?

Using a direct question has to be weighed against an alternative Title that states what you found out. Try different options and see what works best. Using a question can be especially good when you have a short, snappy question, but a complex and cumbersome answer that can't be summarized easily. Questions can also make good titles for review papers, or if you are the first person to ask that question.

But you should use questions sparingly, and only after careful thought. Using a question is risky because it doesn't give the main message of your paper, which is usually the better approach.

Consider using a colon

You can use a colon (:) to help set up a question. For example:

> *Lethal combat in wasps: do males avoid killing brothers or do they never meet them?*
>
> *The evolution of communication in bacteria: when is a signal not a signal?*

You can also use a colon to divide the question from your result (answer):

> *Testing the role of dance in mate choice: females prefer males who dance more symmetrically*

Or to help set up multiple parts of your paper:

> *Foraging behaviour in sharks: testing the roles of weather and prey availability*

A colon can also be used to combine some form of cultural reference or attention-grabbing phrase with the specifics of what you did. For example:

> *Birds of a feather: social networks in parrots*
>
> *Of mice and men: modelling bubonic plague dynamics*
>
> *Cooperate locally, compete globally: plant–fungi networks*

Such Titles can be brilliant if they work and are novel. But there are lots of reasons why they won't work. They may they have been used hundreds of times before, or they may be lost on readers who don't know the book or film you are referring to. Or maybe it just isn't as clever or funny as you think. So use this approach sparingly and test it out on colleagues.

Finally, don't forget that using a colon often leads to longer Titles that can be more cumbersome. So, ask yourself: is that extra length worth it?

Be bold

A Title is a chance to make your paper more striking (Cartoon 9.3). It will grab the attention of your readers and suck them in. Compare:

1. *The sex ratio strategy of* Melittobia *wasps depends upon dispersal status*
2. *A solution to the sex ratio puzzle in* Melittobia *wasps*

Cartoon 9.3 *Be bold!*

Option 1 is a description of the main result and take-home message. But option 2 is more likely to grab the reader's attention. You don't have to know anything about sex ratios, or *Melittobia* wasps, to think: 'Oh, there is a puzzle and they have solved it—I want to know more!'. It's true that option 2 is vaguer, because it doesn't say what was found out, but in this case the benefits could outweigh that cost.

Of course, be careful not to overdo it. It's important not to oversell your paper, so make sure that you deliver on what you have promised. If your Title is 'Can we eradicate human pathogens?' and your actual study looks at the function of a single gene in one pathogen, then you are likely to annoy your readers. They may even be so annoyed that they will avoid your papers in the future. So, be bold within reason!

Use different styles for different types of paper

Different types of paper require different styles. If you have developed a theoretical model to explore a specific question, then that question could work well as the Title. If you have written a review that gives an overview of an area then you might not have a single point to emphasize, and the subject area could work well as the Title. If you are writing a short commentary on a recently published paper, or a book review, then a pun or topical cultural reference could be a good option. Remember: your Title will give an impression of what kind of paper it is, as well as its content.

Don't repeat titles

Looking at the titles of previous papers in your area can be very useful. You are trying to make your paper look interesting and novel enough that people will choose to read it. So don't repeat or almost repeat a title that has been used before. Use your Title to make the novelty of your paper clear, to help convince potential readers to delve into it.

Target search engines

It can be useful to step back even further and think about how people find papers in search engines, such as *Web of Science, Scopus,* or *Google Scholar.* The combination of your paper Title and the keywords that you choose will determine when your paper is found, and by whom. Think carefully about which words or phrases will allow people to find your paper and make sure that they are included somewhere. Again, this means thinking about the kinds of reader that you want to attract. Your Title can help people find your paper, as well as encouraging them to read it.

Sometimes people will even cite a paper without reading it, just because they need a reference for a certain point. A brief Title that makes the message clear can be more likely to be cited in this way. Of course, the aim isn't to just get citations—but a citation can spread your paper to new readers, which is always good.

Summary

Handy Hints for a Terrific Title
Try Different Options
1. Produce about five candidate Titles
2. Read them out loud
3. Leave for a day or two
Guiding Principles
1. Convey your message not your methods
2. Be specific
3. Make the message clear
4. Don't try to do too much
5. Be concise
6. Keep it simple
7. Avoid acronyms and abbreviations
8. There are no hard and fast rules
Fancy Flourishes
1. Consider using a question
2. Consider using a colon
3. Be bold
4. Use different styles for different types of paper
5. Don't repeat Titles
6. Target search engines

Box 9.1 presents a summary example.

Box 9.1 Summary Example

To give a worked example, let's consider some Title options for the hypothetical fruit fly study that we introduced in Box 3.6.

1. *A study of egg-laying behaviour in fruit flies*[2]
2. *An experimental manipulation of social environment shows that foraging fruit flies prefer to lay eggs on bananas where there are other females*[3]
3. *The oviposition behaviour of fruit flies in a heterogeneous social environment*[4]
4. *The scandalous egg-laying behaviour of fruit flies*[5]
5. *How do fruit flies choose where to lay eggs?*[6]
6. *Does the presence of other females influence where fruit flies lay eggs?*[7]
7. *Fruit flies prefer to lay eggs on bananas where there are other females*[8]

We showed a couple of our favourite options to colleagues, who preferred number 7. They found number 6 a bit vague, and thought that giving the actual result worked better. But maybe you disagree, and prefer one of the earlier options, or can come up with a different Title that you prefer. That's fine—there probably isn't a perfect Title on which we can all agree.

---

[2] This Title describes what was done, but it doesn't give the take-home message. The phrase 'A study' doesn't tell us anything—of course it was a study, what else would we be writing up, a recipe? And it is a bit boring—try reading it out and see if it fails the robot voice test (spoiler alert: it does!).

[3] This Title is better, in that it tells us what was found out. It is also interesting to know that it was an experiment, not just an observational study. But it is a bit long (robot test), and the take-home message is a bit hidden among other details.

[4] This Title feels like a step backwards. It is as vague as 1, but also throws in some technical jargon. Many potential readers will have no idea what this study is about.

[5] This Title is over the top and may annoy readers. If you are going to make a big claim, like the existence of a scandal, then it had better really be one. Females preferring to lay eggs where there are already other females is interesting, but it isn't a scandal.

[6] This question sums up a point of the paper, so could work. But we should weigh it up against options that give the main result (message). Also, is it too big a question? The study just looked at one aspect of egg-laying behaviour—the presence of other females.

[7] This is a better, more precise question. It's longer, but gives a much better idea of what was done, so that extra length is justified. At the same time, is it better to give the actual result? Why would this paper need a question for a Title?

[8] This Title is our favourite. It explains the main result—the take-home message. It doesn't mention some other details, such as that females also lay more eggs and spend longer there, but those extra details would make the Title more cumbersome. This Title is a nice, snappy 13 words.

10

Cover Letters

Now that your manuscript is finished, it's time to submit to a journal. In most cases, this is easy: simply upload it to the website of your chosen journal and press submit. But sometimes, the journal requests a cover letter, to supply extra information for the editors—and you can always provide one, even when you aren't asked.

Cover letters play a vital role in getting a paper published. This is particularly true at the highest-impact journals, where editors use cover letters to help decide which papers to send out to review and which to reject immediately. At some journals, this is the biggest hurdle you will face, with up to 80% of papers being rejected without review (Cartoon 10.1).

Consequently, you should be putting the same care and attention into your cover letter that you put into your actual paper. If you don't, you are increasing the probability of a very speedy rejection. The problem is that it isn't always clear what to include in a cover letter. While you will have read many scientific papers, you probably don't have much experience reading cover letters. Key questions include: How much detail is appropriate? How long should a cover letter be?

As in previous chapters, we provide a simple structure for you to follow. This structure should allow you to explain to the editor, as simply and concisely as possible, why they should send your paper out for review.

Write for the Editor

Before discussing how to structure a cover letter, it's useful to step back and think about who will read it. A cover letter is only for the editor of the journal. It is not for all the potential referees and readers. Indeed, journals often explicitly state that the cover letter will not be shared with referees.

Consequently, a cover letter can be laser-focused on convincing the editor that they need to send your paper out for review. It should therefore include a very brief and simple overview of the main message, without complications or caveats. There are lots of things that need to be discussed in papers that can just be ignored in the cover letter! You don't need to convince the editor on every point—you just need to convince them that your paper is of sufficient general interest to the journal's readers.

Scientific Papers Made Easy. Stuart West and Lindsay Turnbull, Oxford University Press. © Stuart West and Lindsay Turnbull (2023).
DOI: 10.1093/oso/9780192862785.003.0010

Cartoon 10.1 *Getting a paper sent out for review can be a major hurdle.*

The type of cover letter will also depend on the journal. If you are submitting your paper to a specialist journal, then the editor will probably know a reasonable amount about your topic. But, if you are submitting to a more interdisciplinary journal, then each editor has to cover a much wider range of subjects, and they may not know too much about yours. It's therefore essential to pitch your cover letter appropriately.

Remember—the editor may know a lot less about the topic of your paper than your potential readers.

Structure your Cover Letter

Journal editors use a cover letter to help assess whether a paper is relevant, important, and of interest to the readers of their journal. A good cover letter will persuade the editor that this is the case. *A good cover letter is a persuasive cover letter.*

One way to write a persuasive cover letter is to structure it into three sections:

1. Summary: *succinctly explain what you have found out.*
2. Justification: *explain why your paper is suitable for this particular journal.*
3. Box-ticking: *state that you have fulfilled any particular requirements of the journal, for example that the data has not been previously published elsewhere.*

This leads to a cover letter that looks something like:

> Dear Editors,
> Please accept our paper '[PAPER TITLE]' for submission to [JOURNAL TITLE].
> We [SUMMARY].
> Our paper is suitable for publication in [JOURNAL TITLE] because [JUSTIFICATION].
> Our paper [STATEMENTS REQUIRED BY JOURNAL].
> Yours,
> [YOUR NAME]

We will now explain how to fill in the details.

Summarize your paper

A cover letter begins with a summary of the main finding or take-home message. Remember that you don't need to summarize everything; instead, focus on one or two things of greatest interest to the readers of the journal (and hence the editor). We provide several examples of summaries, because different formats work well for different papers (and journals).

Example 1 is a very simple statement of your results:

> *On average, we found that 11 different viruses make the jump from domestic animals to humans every year.*

Example 2 is a longer statement that includes information about the methods used, because they were especially novel or impressive:

> *We collected data on 5,469 species of mammal and found that carnivorous species were three times more likely to be at a very high risk of extinction, relative to herbivores and omnivores.*

Example 3 is a short paragraph that includes some context, to explain why the results matter. This could be introductory material to set up your study, and/or its implications:

> *Organic farming, which avoids the use of insecticides, herbicides, and chemical fertilizers, is generally thought to provide a better habitat for wildlife. However, there has been no global assessment of whether organic farming leads to an increase in the diversity or density of wildlife. We collated data from 73 studies, in 27 countries, which had examined the influence of farming method on birds. We found that organic farming leads to a 30% increase in the number of bird species observed on a farm and a 75% increase in overall numbers. These results provide a clear demonstration of the environmental benefits of organic farming.*

Example 4 is a slightly longer summary that includes more than one paragraph. This is similar to Example 3, and also includes information on the background, methods, results, and implications, but in this case we have separated the information into three different paragraphs, which can be easier to read:

> *The last 20 years has seen a revolution in our understanding of how fungi and plants trade resources, in a form analogous to an economic market. Experiments have shown*

that the amount of carbon that plants provide to fungi depends on the amount of phosphorous that they obtain from the fungi, and vice versa.

However, these advances have been driven by laboratory experiments in relatively unnatural conditions. There are no data that demonstrate the importance of such resource trading in natural communities.

We used quantum dot labelling of phosphorous and carbon to follow patterns of trade in a natural grassland. We found that plants obtain an average of 85% of their phosphorous through trading with fungi, demonstrating the key role of this trade in nature.

To work out which format is best for your paper, first determine the most important points and then play with different possible formats. Think about your target editor, and what it would take to persuade them. Remember that a longer summary isn't always better. Editors, like reviewers and readers, are busy people.

Don't just cut and paste the Abstract

The summary of your cover letter is likely to contain similar information to the Abstract of your paper. However, do not just cut and paste the Abstract into your cover letter (Cartoon 10.2). The editor will already have your Abstract, so if you repeat it you are wasting an opportunity to summarize your work in a different way. In particular, when deciding on the format of the summary in a cover letter, you have a free hand, whereas an Abstract must follow the rules of the journal.

In addition, don't forget that the Abstract and the cover letter are aimed at different audiences. The Abstract is aimed at researchers, who have chosen to read it, probably because they liked your Title and know that your paper is relevant to their work. In contrast, the cover letter is aimed at the editor assigned to your paper. The editor has not chosen the paper and may have little detailed knowledge of the subject. Consequently, a cover letter will usually need to be simpler, and focused on the points of broadest interest.

Remember—a cover letter just needs to convince the editor that there is something interesting in the paper; it doesn't need to provide a technical summary.

Justify the suitability for the journal

When considering a new submission, the editor only has to ask themselves one question: 'Why should we consider publishing this paper in our journal?' There is only one general answer to this question—because the journal's readers will want to read it. So, by making it clear why those readers will want to read your paper, you can justify to the editor why they should send it out for review.

Before we consider some specific examples, let's consider who you might be targeting and why. The potential readers of your paper range from specialist to generalist, and might even include non-scientists, such as policymakers. So, to help you think about who

Cartoon 10.2 *Don't just cut and paste the Abstract!*

your audience might be, we have provided five questions that can help you formulate who will be interested and why.

1. What is the main implication of your work for other specialists in your field?
2. Does your paper develop a methodology or provide a data set which other researchers in the same or similar fields could use?
3. Does your paper have ramifications for other fields or shed light on a larger issue?
4. Are there any applied implications of your work? For example, does your work suggest action that needs to be taken in the real world?
5. What is the widest group of readers that could be interested in some aspect of your paper? This could include people in entirely different disciplines, policymakers or even the general public.

By considering the answers to these questions, you will find it easier to come up with a number of reasons why your paper is important to more than one potential audience. It's important to remember that you might not have answers to all of these questions, especially if you are submitting your work to a more specialist journal. Indeed, don't overstretch your case. If your paper is on mate choice in fruit flies, don't pretend that it's directly relevant to government officials.

One way to structure your justification section is to actually write a numbered list. If you choose to do so, then keep it short and hard-hitting. A list of three to five reasons indicating why your paper is suitable for a particular journal will usually be enough. Indeed, a longer list could harm the chances of your paper being sent out for review, because less important points will inevitably dilute your main message. We now provide some examples.

Example 1 is a single brief statement:

> *Our paper is suitable for [JOURNAL NAME] because it provides urgently needed data on the risk that domestic animal viruses pose to humans.*

Example 2 is a longer statement that includes implications of the findings for conservation practitioners and policymakers:

> *Our paper is suitable for [JOURNAL NAME] because it identifies a single factor with a large influence on whether mammal species are currently listed as endangered by the International Union for Conservation of Nature. This provides a framework for determining which species urgently need more protection.*

Example 3 is a numbered list that makes it clear how the paper will be relevant for different groups:

> *Our paper is suitable for [JOURNAL NAME] because:*

> 1. *We provide the first quantification of the importance of resource trading between fungi and plants in a natural community.*
> 2. *We develop a novel methodology for tracking resources with quantum dots, which could be applied to a wide range of study systems.*
> 3. *Our results help to explain how cooperation between species can be favoured in the natural world.*
> 4. *The topics of economic trade and cooperation are of broad interest across the sciences, mathematics, social sciences, and to the general public.*

Write a Persuasive Cover Letter

The above sections have outlined the basic structure for a cover letter. We now provide some simple tricks for filling out this structure, to make your cover letter even more persuasive, and hence maximize the chance that it will be sent out for review (Cartoon 10.3).

Start early

A cover letter is crucial in getting your paper sent out for review. Consequently, do not rush the cover letter or leave it to the last minute. It's a good idea to start working on the cover letter at the same time as the Abstract. They both involve summarizing your paper, so it can be efficient to write them at the same time, but you should ensure that they are complementary and don't just repeat the same material. Another option is to start working on a cover letter when the draft manuscript is with co-authors or colleagues for comments.

Once you have written your cover letter, treat it like any other part of your manuscript. Leave it for a few days and then come back to edit it, when you have 'fresh eyes'. Get

Cartoon 10.3 *Write a persuasive Cover Letter.*

co-authors and colleagues to read it and give comments. As with any other section of your paper, you should edit and polish your first draft.

Be concise

There is always an advantage to being succinct. Remember that the cover letter isn't a technical summary of your paper, so you don't need to explain all the fiddly details—if the editor is interested, then they can look them up in your actual paper. Instead, focus on the general interest of your paper and think about the big picture. What would readers of that journal be most interested to find out?

While there are no strict rules, we suggest that you should keep the summary and the justification to no more than 150 words each. By constraining yourself, you will be forced to focus on the major points in a form that can be easily digested by the editor. Of course, some cover letters will need to be longer, especially if you are summarizing a complex paper or you have written a paper on a complex issue. But 150 words is a sense check. If one or both of your sections are longer than this, then stop and think about whether you can really justify the extra words.

Sometimes editors will decide as a team which papers to send out for review. In this case, one editor may have to summarize your paper and explain its virtues to the others. If you have provided a concise and clearly worded summary, then you have made their job easier—and this means that your paper is more likely to be sent for review.

Consider using subheadings

If you feel that your cover letter is too long, you could consider breaking it up with subheadings. For example, you could put *Summary* before the summary of your paper, and *Why [JOURNAL NAME]?* before your justification.

Consider using a figure

Figures are not normally put into cover letters. Consequently, adding a figure could look strange to an editor. But, if a figure would really help to make things easier and reduce the text needed, then it could be justified. The figure could be a key result or even a Graphical Abstract (Chapter 8).

Try different options

A cover letter should be quite short, so it shouldn't take long to plan different options and see what works best. For example, try putting the stress on different points, including a different number of points, or putting the points in a different order.

Suggest referees

Some journals request suggestions for potential referees. An editor will usually add these to their own list, and then go through them until enough people have agreed to review the paper. Editors often have trouble finding enough reviewers, so suggestions are very welcome. So who should you suggest? We have a few key pieces of advice.

First, suggest reviewers that the editor might easily overlook. The editor's list of potential referees will often be drawn from the citations in your paper, or researchers that they already know. Your suggestions will be most useful if they don't overlap with the people that the editor would choose anyway, so think hard about less obvious choices. Perhaps there is someone who does good research in a related area, but who you haven't cited in your paper. Or perhaps you know of a great early-career researcher who is moving into your field, but who the editor may not have heard about or considered.

Second, suggest people who you actually *want* to referee your paper. This might seem obvious, but reviewers often give really useful feedback, even when they don't recommend immediate acceptance! And this feedback can help you produce an even better paper.

Third, suggest people that are more likely to agree when asked. There are a few things to consider here, including the current interests of your potential reviewers and their career stage. Someone who is still actively working in your subject area is likely to be more interested in investing the necessary time than someone who is now working in another field, while people at an earlier career stage might be less busy than the established names in your field. Indeed, the established 'big names' often receive more invitations to review than they can possibly manage, so they are more likely to say no.

Fourth, when choosing your suggestions, think broad and balanced. Are you thinking internationally, rather than just considering people from your own and neighbouring countries? Are your suggestions gender biased?

Finally, don't make too many suggestions. An editor will likely pick one, but only one, of your suggestions. Consequently, just suggest two or three potential referees. There is also no point suggesting people that the editor will not or cannot choose, for example colleagues that you have recently published with, or members of your own institution.

When suggesting referees, remember to include the reason why you are suggesting them:

As potential referees, we suggest:

1. *Professor Jean Grey (New University, USA;* jean.grey@new.edu*)—International expert on genetic manipulations.*

2. *Dr Anna Lebeau (Old University, France;* anna.lebeau@old.fr*)—Early-career researcher with expertise at the cutting edge of proteomics (see publication X).*

We recommend adding referee suggestions to all cover letters, even if the journal doesn't request them. Referee suggestions will often be useful for the editor, and this is your chance to influence who reviews your paper. Making referee suggestions takes very little time, so you haven't lost much, even if the editor doesn't use them.

You can also request that someone should *not* be used as a referee. This should be done sparingly and only when there is a very good reason. For example, good reasons to request that someone should not review your paper include strong conflicts of interest or a personal dispute. They do not include simply thinking that a certain person might not like your paper.

Write a Persuasive Cover Letter, Even When You Don't Have to

Some journals request only a very minimal cover letter. For example, one that just includes statements about previous publication and conflicts of interests. Other journals do not request a cover letter at all. However, it can still be worth writing a cover letter that summarizes your findings and states why your paper is suitable for the journal. If there is any chance that an editor will reject your paper without sending it out for review, then it can be worth writing a persuasive cover letter to reduce the probability (and hurt) of that instant rejection.

Summary

Great Guidelines for Creative Cover Letters
Write for the editor
1. Convince the editor, not referees 2. Pitch appropriately for the journal
Structure your cover letter
1. Summarize your paper 2. Don't just cut and paste the Abstract 3. Justify the suitability for the journal
Write a persuasive cover letter
1. Start early 2. Be concise 3. Consider using subheadings 4. Consider using a figure 5. Try different options 6. Suggest referees
Write a persuasive cover letter even when you don't have to

Box 10.1 presents a summary example.

Box 10.1 Summary Example

In this box we give two versions of a cover letter for the hypothetical fruit-fly foraging research that we have been examining throughout this book. The first version has problems, which we have highlighted and annotated. The second version is based around the framework and tips from this chapter.

Version 1

> *Dear Editors,*
>   *Please consider our paper, 'Fruit flies prefer to lay eggs on bananas where there are other females', as a research article for* Animal Behaviour.[1]
>   *Our results have implications[2] for a wide variety of disciplines, and thus will be of broad interest.[3] Entomologists and other animal behaviour workers will be interested in*

---

[1] This is a relatively general journal, which publishes on many aspects of behaviour across a range of organisms. Consequently, the editor is likely to be a non-specialist.

[2] This cover letter opens with a discussion of why the paper is important (its implications), before it has explained the results. The editor will therefore not be in a good position to follow or assess the implications.

[3] It is better to *explain* why the paper is of broad interest, rather than just *say* it is of broad interest.

how we have experimentally manipulated the social environment.[4] A lot of energy in animal behaviour has been devoted towards understanding foraging. Our results suggest a novel approach.[5]

We[6] conducted clear and powerful[7] experimental manipulations to examine the role of a potentially important factor in the foraging behaviour of a fruit fly.[8] Contrary to previous studies,[9] our results allowed us to examine whether females adjusted their oviposition[10] behaviour in response to the presence of other females.[11] This[12] allowed us to assess the role of a factor that may potentially be important in explaining the economic impact of this fruit fly.[13]

We found the important[14] result that females adjusted their oviposition behaviour in response to the presence of other females.[15] The consequences of the social environment had not been investigated previously,[16] and the possible role of this factor was not mentioned in a pivotal review of fruit fly behaviour, that has been cited over 1,000 times (Fisher & Fair, 2002).[17] Although field observations had suggested the possibility of aggregation, there had been no direct experimental test of competing explanations.[18] Our results provide a novel result about foraging behaviour, and also could be exploited as part of a fruit-fly mitigation

---

[4] The experimental methods were relatively standard—probably not worth emphasizing as one of the major points.

[5] This paragraph is vague and not very punchy.

[6] This is quite a long cover letter—subheadings could have been useful. Subheadings might have also helped (or forced) the authors to use a better structure, for example by separating the summary and the justification, rather than bouncing back and forth between them.

[7] The meaning of 'clear and powerful' isn't clear. Most experiments should give relatively clear results.

[8] It hasn't been explained why we should care about foraging in a fruit fly. Is it to test foraging theory? Is it because fruit flies are important?

[9] Were the results contrary to previous work or just testing a new factor?

[10] Jargon like 'oviposition' is likely to be even tougher for an editor than it would be for a potential reader.

[11] Quite a vague sentence—trying to sound grand, without really saying much.

[12] Not clear what 'This' is referring to.

[13] This sentence provides context for why the experiment was done, after the result has been explained. It would have been better to give this context first. In addition, the explanation could be a lot simpler, to make sure it could be understood by a non-specialist editor. Finally, not enough has been explained for the editor to understand why and in what way this paper might have economic impact.

[14] Don't say that a result is important—explain why it is important.

[15] Doesn't give the direction of the result.

[16] This sentence is vague and uses terminology that an editor might not understand. In addition, the direction of the result still hasn't been given.

[17] This complaint about omission in a previous review doesn't really help. It comes across as a bit of a whine, and the review is quite old. Presumably the field has moved on since then. What is the state of the art?

[18] It isn't clear what this previous work is, or its limitations, which might leave the editor thinking: 'maybe the problem was solved before this study'. It is not the job of the cover letter to link to all previous studies. In addition, this is providing background after results.

scheme,[19] *and*[20] *we carried out a number of different analyses to test the robustness of our experimental data.*[21]

In summary, we argue that this paper is suitable for Animal Behaviour *because:*[22]

1. *It provides a decisive result for animal foraging, that also has agricultural implications, and so will be of widespread interest (*Animal Behaviour, Evolutionary Ecology & Agronomy*). Consequently, it will be of direct interest to a particularly high number of researchers.*[23]

2. *It is the result of an elegant yet powerful experiment, designed to mimic natural conditions.*[24]

3. *The subject of pest management is economically important.*[25]

This work has not been published elsewhere, and the authors approve of the submission. Yours, Roberta Hood[26]

Version 2

Dear Editors,

   Please consider our paper, 'Fruit flies prefer to lay eggs on bananas where there are other females', as a research article for Animal Behaviour.

Summary[27]

Fruit flies are the most important agricultural pests of soft-skinned fruit. The adults lay eggs on fruits, which the larvae then emerge on and consume.[28] *The damage caused by fruit flies and our ability to reduce this damage depends on how females choose where to lay their eggs. We carried out experiments in flight cages to examine the foraging behaviour of* Belial lys, *a major pest of bananas. We found that females prefer to lay eggs on bananas where another female was already present (aggregation). This finding offers a potential route for new control methods.*

---

[19] This sentence is perhaps better left for the justification section.
[20] This sentence addresses two very different issues and should be split into two sentences.
[21] This sentence is returning to methods after results, conclusions, and implications. In addition, while it can be good to emphasize the robustness of results, this is taking up space, and is not one of the most important points, as the experiment was relatively simple.
[22] This paragraph feels like the justification section is starting again, which will confuse and/or frustrate the editor. It is good to list the key points, but why were some things in a paragraph, and others in a list? In addition, a cover letter is supposed to *be* a summary—it shouldn't need its own summary!
[23] Not the best phrasing, although it might be true.
[24] The results are more important than the methods.
[25] It feels like the applied implications are being pushed too much. This is an animal behaviour study, which has applied implications, but they need to be independently developed.
[26] Version 1 of the cover letter is very long—over 1,000 words!
[27] The summary of the paper is approximately 100 words.
[28] This text adds extra context (background biology) relative to the abstract, which an editor might not know.

Why Animal Behaviour?[29]

Our paper is suitable for Animal Behaviour *because:*

1. *We provide the first experimental demonstration that fruit flies adjust their choice of egg laying sites in response to the presence of other females.*

2. *The influence of the social environment on foraging behaviour is a major question in the study of animal behaviour across all taxa—our results suggest positive interactions in an insect.*

3. *Our results suggest a novel factor that could be exploited[30] in the production of scent-based traps or lures for fruit flies, to reduce their economic cost to agriculture.[31]*

This work has not been published elsewhere, and the authors approve of the submission.
Yours,
Roberta Hood[32]

Comparison

Version 1 is badly structured and too long. The key points were not laid out clearly to the editor, and the text jumps back and forth between the summary and the justification. Version 2 is better structured and much shorter. Conclusion? The major points jump out much more clearly in the second version, and the same paper has a much better chance of being sent out for review.

---

[29] The justification section is approximately 100 words.
[30] The details of the applied significance are better left to this section.
[31] Note how this list begins with the specific result and ends with the broad implications.
[32] Version 2 of the cover letter is less than 300 words long.

11

Writing and Editing

In the previous chapters we have provided a structure for producing each section of your paper. In this final chapter we consider how to write and edit in an effective and efficient way. This includes how to divide your time among the different sections, so that your paper emerges efficiently, and how to know when your paper is ready to submit.

Following our tips for efficient and effective writing will allow you to produce a first draft more quickly. Then, once you have finished that draft, you will need to edit it. Editing is much more than just a final polish—it can play a major role in ensuring that your paper is readable and will have impact.

How to Write?

There are many decisions about how to write a paper, from which section to write first to how you actually sit down and write. We provide some tips that we think are helpful, but the most crucial thing is to find out what works best for you, which means actively trying different options—including ones that you might want to dismiss at first glance.

Where to start?

When considering the order in which to write the different sections of your paper, one option is to follow the order of the chapters in this book: Methods, Results, Introduction, Discussion, and Abstract (Cartoon 11.1). The advantages of writing in this order are:

- You begin with the sections that are easiest to write. Getting these done can make writing the rest of the paper less daunting.
- You begin with the core of your research—what you did and what you found out. It can be very useful to get this firmly sorted in your head before proceeding with other parts of your paper.
- The Methods can be the most boring section of the paper to write, so it is great to get it out of the way!

Scientific Papers Made Easy. Stuart West and Lindsay Turnbull, Oxford University Press. © Stuart West and Lindsay Turnbull (2023).
DOI: 10.1093/oso/9780192862785.003.0011

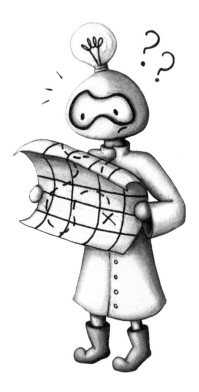

Cartoon 11.1 *Where to start?*

Another popular option is to start with the Introduction and then write the sections in the order that they appear in the paper, but leaving the Abstract to the end. This means writing sections in the order: Introduction, Methods, Results, Discussion, Abstract. Potential advantages of starting with the Introduction include getting a possibly tough section out the way and being forced to think hard about why you did the experiment and which questions you were actually trying to answer! Writing an Introduction will force you to step back and think about the big picture. It can even be useful to write the Introduction before you start your statistical analyses!

A third option is to start by writing the Abstract, and going on through the paper in the order that the sections will appear: Abstract, Introduction, Methods, Results, Discussion. One advantage of this approach is that you are mimicking how a reader will see the sections, and so can perhaps better imagine what they need. However, the Abstract is generally difficult to write, especially for less-experienced writers.

Finally, you could choose a different sequence that you think makes better sense! Our chapters are stand-alone, so you can just read them as required.

Whichever order you choose to write your sections in, it can be useful to make notes for other sections as you go along. Writing the Methods section might make you aware of methodological weaknesses that you want to raise in the Discussion, while writing the Results might flag issues that need to be included in the Introduction. Indeed, the different sections must feel coherent—and this means that you will inevitably have to edit each section again, once further sections have been written.

When to start?

One option is to only start writing your paper once all of your research is finished. This is probably the most common approach and does make sense—after all, it's only once everything is finished that you have a complete set of results on which to base your story. Indeed, many people have already started new projects before getting round to writing up their work, although it's a good idea not to leave too long a gap between doing work and writing it up.

However, it can be useful to write up some bits as you go along, especially the Methods and Results sections. You could write up your Methods as you are carrying out your research, or immediately afterwards, before you start analysing your data. At this stage, you will know exactly what you did and any problems that came up, and you are more likely to remember all those essential details.

You can also write up your Results as you do the analyses. Again, your understanding of your analyses is likely to be highest while doing them, so even if you don't feel in a position to formally start writing, some really good notes and/or 'key sentences' will make your life a lot easier when you do eventually find time. This is also a great time to finalize potential figures and to share them with co-authors. Some people like to produce figures to help plan their Results section.

As well as the actual writing of your paper, it can be useful to annotate your analyses or code as you go along. For example, code for statistical analysis or programming, or notebooks for mathematical analyses. This can serve two purposes: as notes for yourself, but also as part of your publication. Many journals now require authors to include their data and code as supplementary material online, as part of a drive for reproducible research.

The same principles apply to theoretical modelling, where it can also be useful to draft Methods and Results sections as you go along.

How to start?

In Chapter 2, we emphasized the usefulness of planning, rather than just sitting down to start writing (page 9). Good planning helps you to construct a better paper and can make things a lot easier in the long run. But as well as planning your paper, it can be useful to actively plan how to write it.

Writing is best achieved with good-sized chunks of time without interruptions. Make sure you have time to think, and don't feel rushed. Think about where you can best concentrate and are least likely to be interrupted. This might be your office, your home, or even a favourite coffee shop. It can even aid concentration to move around between several different locations. Think about whether you find the presence of other people distracting or not—and if you do, but don't have a quiet space available, consider investing in a decent pair of headphones.

As well as needing some proper quiet time, most people's productivity can be improved by taking short breaks. Concentration inevitably drifts over time, and it can be useful to take a break to reset and gather your thoughts. Having regular breaks can help you achieve more in the long run, so don't slump over your computer for hours at a time, thinking that this is the only way to get your paper written!

The 'Pomodoro Technique' is one way to enforce breaks.[1] It is named after a tomato-shaped kitchen timer—pomodoro is the Italian word for tomato. The technique recommends writing in 25-minute periods, called pomodoros. After each pomodoro, you take a short 3–5-minute break, to pause, stretch, or make a cup of tea. After four pomodoros, you take a longer 15–30-minute break, which might allow you to take a walk or even do some other form of exercise, such as yoga. It's important for your physical wellbeing to move around during these breaks. Some academics suffer from postural conditions because they have spent too long sitting at their computers for extended periods of time. Timing your pomodoros can be managed with a range of modern apps, but there can be a psychological benefit to winding a physical timer.

There are many variations to the Pomodoro Technique and you can tailor the details to suit your writing style. The crucial point is to think about how to incorporate breaks into your writing routine, to improve effectiveness and to keep yourself healthy. Some people even find it helps to write in a group or 'club'.

You may also want to consider which journal you are targeting. Some people do this right from the start, while others prefer to write a more 'generic' paper that can then be tailored for specific journals. If you are targeting a specific journal from the start, then don't forget to check their Instructions for Authors. While some journals are relatively flexible, others have specific restrictions over issues such as the nature and length of each paper section. It is essential to follow these instructions, unless you want to give an editor an excuse for an instant reject.

Edit, Edit, Edit!

Nobody—and we mean nobody—writes a perfect draft with their first attempt. Thankfully, this isn't a problem when writing scientific papers, as unlike writing a news article, you don't have to finish by the end of the day. Revisit your work, and *edit, edit, edit!*

It is surprisingly easy to lose track of the big picture, get lost in the details, or not follow all the tips that make for better writing. We commonly write sentences or sections that contradict our own advice. Editing is the answer to these problems, allowing you to polish up your paper into something that a reader will cheerfully read.

Unfortunately, editing is hard and there aren't any short cuts. If you look back at something you recently wrote, it's almost impossible to be objective. Rather than actively

---

[1] See https://francescocirillo.com/pages/pomodoro-technique.

reading what you wrote, it's easy to slip back into your previous thinking about that section, replaying your 'thinking voice' in your head. Your eyes glaze over and you can miss blatant errors, as we are sure to have done in this book. It's especially difficult to step back and assess the details that you really need to explain to the reader—this is the 'curse of knowledge' problem we mentioned in Chapter 2. We suggest five routes to better editing.

Leave and forget

Leave something for a few days or a week before coming back to it. Even a short break can allow you to come back with fresh eyes, and spot glaring problems that you would otherwise have missed.

Be forensic

Actively try to pull apart what you have written (Cartoon 11.2). Editing also provides a great opportunity to check that you have followed all our tips. Have you structured your

Cartoon 11.2 *Be forensic.*

Methods in the best way? Have you put why before how? Have you kept it simple, with short sentences? Have you used terminology consistently?

You need to actively work through what you have written, examining at all levels, from the overall structure to the details of each sentence. Apply the tips and principles from this book to every paragraph, and every sentence. We know that this sounds like hard work, and it is. But it is a crucial part of the scientific process. There is no substitute. Remember that no matter how amazing your research is, it will only have impact if people read and understand it. If you produce a clear and easy-to-understand paper, this will make a difference to every person who tries to read it.

Remove unnecessary words/sentences/paragraphs

Good writing is concise. The shorter it is, the easier it can be for the reader to take in the key points. However, it can be hard to produce a concise version first time around. So, start with a longer version and accept that you're going to have to commit to several rounds of editing.

Editing gives you a chance to trim out unnecessary words and sentences. Be ruthless—you are likely to be surprised at how much you can trim by actively editing your manuscript, without losing any meaning. Here are some examples:

1. *(See figure 3)*.
2. *(Figure 3)*

Why else would you mention a figure or a citation?

1. *Here, we analyse . . .*
2. *We analyse . . .*

Where else would you be doing it, but here?

1. *We see that the . . .*
2. *The . . .*

1. *We note that the . . .*
2. *The . . .*

1. *We repeated this procedure for six iterations (rounds).*
2. *We repeated this procedure six times.*

1. *Research in this area has . . .*
2. *Research has . . .*

1. *It has long been known that the . . .*
2. *The . . .*

1. *There is a chance that moths will be able to . . .*
2. *Moths could . . .*

1. *It is clear that to gain further insights into the factors that determine gene regulation, we need to measure how . . .*
2. *Further insight requires data measuring how . . .*

1. *In microbial systems, intraspecific competition is induced by stressful conditions such as resource or space limitation.*
2. *Bacteria compete for space and resources.*

1. *We listed all the information regarding the full genome analysis of the strains used in this competition assay in Table 1.*
2. *We listed all mutations in Table 1.*

Don't stop at words. Sometimes you can easily remove whole sentences.

1. *Our results show that more work is required in this area.*
2. *[Deleted]*

Sentences like this don't really say anything—more work is always needed—so can be just deleted.

1. *These results are discussed.*
2. *[Deleted]*

Of course they are, you don't need to say that!

1. *This relationship is discussed in figure 3.*
2. *[Deleted]*

Instead, just put *(Figure 3)* at the end of the sentence describing the relationship.

1. *Understanding human behaviour presents a grand challenge across disciplines.*
2. *[Deleted]*

Sometimes attempts to start abstracts or introductions can produce sentences that don't really say anything. They can also betray a lack of understanding or confidence.

You might even be able to completely delete a paragraph. For example, this is a final paragraph from a Discussion:

> *Overall, we found that bird species varied in the number of eggs that they laid, and that the lifestyle, rather than climate, is more important in determining the number of eggs. We have shown the importance of controlling for confounding factors in statistical analyses, and the need for substantial sample sizes. Although there is large variation across species, this is likely to be genetically controlled and under natural selection. It is clear that to gain further insights into the important factors, further data is required.*

This paragraph is a mixture of sentences from earlier in the Discussion or so obvious that they don't say anything. Delete!

The final paragraph of the Discussion can easily end up not really say anything, and so can be a good candidate for deletion, or at least some serious reducing. Another possibility for deletion is a paragraph that isn't appropriate for that section—for example, a paragraph of discussion in the Results section, or a paragraph of introduction at the start of the Dicussion section.

We use the verb 'to strunk' to refer to editing that removes needless words. Strunk and White emphasized the importance of omitting needless words in their book on writing, *The Elements of Style* (Pearson 1999). It is a good challenge to go through a manuscript and see how much you can strunk it. Can you strunk on 10% of the lines? Can you strunk 5% of the words? Strunking can be especially useful when there are word limits to sections—often it is easier to retain the meaning by strunking words from sentences, rather than removing sentences.

But don't forget to ask yourself: have I removed too much? Is it still readable? Don't go too far—you want to make it as concise as possible to help, not hinder, readability.

Maintain the narrative flow

Editing is not just about correcting errors and improving sentences. Editing offers an opportunity for checking the narrative flow. How well does your paper tell the story of your science? Step back and examine how the different paragraphs link up, both within and across the different sections. One way to do this is to read just the first sentence from each paragraph. If you link these sentences together, are you leading a reader through all the key points, in the right order?

A word of warning—brutal editing can sometimes destroy the narrative flow of a paper. It is your job to ensure that links and consistency between the different parts of the paper are maintained. This is especially true when dealing with edits by co-authors, comments from colleagues, or responses to referees. Indeed, different edits or comments can contradict each other, pulling you in different directions. Your job is to take responsibility, and address the issues raised, while maintaining the narrative flow. You don't want a disjointed paper that looks like it was written by a committee.

Learn to love critics

Give your paper to a friend or colleague for comments, and explain to them that you want brutally honest comments. Ask them to highlight anything that didn't make sense. When some people give comments, they focus on language corrections. While that is useful, it can be even more useful to get critical comments like: 'this paragraph made no sense' or: 'I had no idea what you have done here'. Find out who gives such critical comments and use them, or explain to a friend that those are the sort of comments you want. Comments that raise major issues about structure and understanding can be a huge help when revising a paper.

It can be very painful to get harsh or extensive comments on your work. Writing a paper takes time and effort, and everyone just wants to be told: 'yes, that is great, job done'. Instead, extensive comments can leave you feeling that you are useless, have done a bad job, and now have a mountain to climb. But if you can step back, a more accurate perspective is that receiving extensive comments is just a standard part of writing a paper. Writing is hard work, taking multiple rounds of revision. Comments that are tough to receive are useful because they speed up the revision process, and help you produce a better paper. Prepare yourself mentally to receive them (Cartoon 11.3).

Cartoon 11.3 *Learn to love critics.*

Comments on a paper draft can be especially useful for solving the 'curse of knowledge' problem that we discussed in Chapter 2. It is very hard to step back and write in a way that makes everything easy for the reader. Comments that feel harsh are usually just pointing out when you haven't been able to do that. We cannot stress enough that everyone gets tough, painful comments, even those that think they can write a book about paper writing! The good news is that, as you get more experienced and work at your writing, you will be better prepared for painful comments, and you may even get less of them.

One way to deal with painful comments is to read them, and then leave them for a couple of days. It's natural to be upset at first, but time is a great healer and can help to mellow your initial response. Instead of wanting to cry or throttle the person that gave you the comments, you come back thinking: 'yes, that is reasonable, those changes would really improve my paper'. Comments can even be useful when you don't agree with them. A colleague might suggest changes that you think wouldn't work, but their suggested changes alert you to an area that needs greater clarity.

It can be hard to work out when is the best time to give a draft to someone else for comments. The more you have worked on it, the more useful feedback will be. However, there is no point spending too long on perfecting a sentence or paragraph that someone then just suggests that you delete. A good rule of thumb is to wait until you find yourself only making minimal changes, and perhaps even going backwards and forwards over the same points. At this stage it's probably time to give it someone else.

When to ask for comments can also depend upon who you are asking. If you are asking a co-author, then it is fine to ask earlier. Indeed, it's a good idea to send them a rough outline and discuss the structure, before you start the detailed writing. You could also send a co-author one section of your paper for comments; for example, you might want to get the Methods section sorted before going onto the Results, and so on.

Use whatever trick works

Different writers use different tricks to help them read their manuscript with fresh eyes. It is worth trying different options, to see what works for you. You could print it out and work on a hard copy with a pencil rather than on your computer screen. You could read it out loud or get your computer to read it out loud. You could change the font to something jarring, like Comic Sans. You could read it backwards, paragraph by paragraph, or sentence by sentence. Tricks like these can be useful for helping spot problems. Although, they are usually not as good as getting someone else to read it, or just setting the manuscript aside for a week or two.

In Box 11.1, we suggest some extra things to think about as you edit, to make things even easier for the reader. Bonus tips!

Box 11.1 Bonus Tips

Once you have your basic structure sorted, there are some small tricks that you can use.

1. **The rule of three (and no more!)**

It is well known that people are very receptive to things coming in threes (tricolons). For example:

> *I came, I saw, I conquered.*
> *Liberté, égalité, fraternité!*
> *Faster, higher, stronger.*

This 'rule of three' is common in literature, slogans, and catchphrases, but can also be used to help convey information in scientific writing. The rule of three can be used with words or phrases that have related meanings, be they adjectives or nouns:

> *The current methods are long, difficult, and expensive.*

Most common definitions are fluid, unclear, and terms are used interchangeably.

Plants produce hairs, spines, and thorns to defend themselves against herbivorous insects.

The forest floor is dark and gloomy and receives on average only 1% of the light incident on the canopy.

The rule of three can also be applied to examples or ideas:

Biodiversity is important for conservation, human wellbeing, and the continued functioning of ecosystems.

Species might coexist because they use different resources, are attacked by unique pests, or respond differently to the physical environment.

Geneticists use Arabidopsis thaliana because it grows fast, is self-compatible, and completes its life cycle in 28 days.

Putting something into a three can make a strong and clear point, getting information across in a concise, efficient, and convincing way. But you shouldn't use them too often, or they will lose their impact, and you might tire your reader out. Save your threes for when you are making important points that you want to hammer home.

More standard sentences can use pairs. For example, pairs of related adjectives or ideas:

Leaves are often tough and well-defended against insects.

Mycorrhizae have an intimate and mutualistic relationship with their host plants.

Tree rings can be used to measure the response of species to rainfall and temperature.

The previous methods were complicated and difficult.

The model is simple and easy to understand.

Arabidopsis is much loved and used by geneticists around the world.

Putting ideas into a pair like this gives a sense of completeness to a sentence, so the reader is not necessarily expecting any further clarification. Very often using a single example or word leaves almost a question mark hanging. For example, simply saying: *The previous methods were complicated* or *The model is simple* is crying out for an explanation of why.

Similarly, two related facts can easily be accommodated within a single sentence. Often this produces a simple yet satisfactory rhythm to a paragraph that's very easy to follow, even if it's somewhat uninspiring. For example:

We arranged the pots on a table in the glasshouse and re-randomized them regularly. We watered the plants every day and applied fertilizer once a week. Finally, we harvested all above-ground parts and weighed them on a microbalance.

2.   Use parentheses carefully

Parentheses (brackets) can be very useful, for example to add in alternative or technical wording. For example:

> *In many ant species, workers specialize on performing different tasks (division of labour).*
>
> *Bacterial toxin production is controlled by a regulatory system that detects DNA damage (the SOS response).*
>
> *A key parameter is the extent to which parasites increase the mortality rate of their hosts (virulence).*

This allows you to write in simple language, without jargon, but still include the jargon in parentheses, to help link to the relevant literature. If you write the jargon first, and then the explanation in parentheses, the reader can get confused and frustrated, as they will read the jargon before they find out what it means.

However, writers also use parentheses to add in extra clauses or facts. This can break up the flow or chain of thought in a sentence, making it clunky and harder to read. Instead of using parentheses, you could include everything in one sentence, or use multiple sentences. If you need to use parentheses, it is often better to put them at the end of the sentence. Compare:

1. *Studies from several countries (which measured mental health and antidepressant use) have suggested that the global financial crisis of 2008 caused (as well as the economic consequences) a mental health crisis.*

2. *Along with its economic consequences, data from several countries suggested that the global financial crisis of 2008 caused a mental health crisis, with an increase in both mental health problems and antidepressant use.*

1. *We found that male birds that did not obtain mates (N = 103 non reproductive males) showed significantly higher levels of the stress hormone corticol-b (t = 13.3, p < 0.01) compared to males which had mated a female (N = 56 reproductive males).*

2. *We found that male birds that did not obtain mates showed significantly higher levels of the stress hormone corticol-b compared to males which had mated a female (N = 56 reproductive males; N = 103 non-reproductive males; t = 13.3, p < 0.01).*

1. *If we assume participants perfectly understand the public-good game (as is often done by economists) then participants will not cooperate (contribute to the public good) when playing with computers (regardless of the costs involved).*

2. *Economists often assume that participants perfectly understand the public-good game. They therefore assume that participants will not contribute (cooperate) when playing with computers, regardless of the costs involved.*

In the examples above, the first version has one or more parentheses breaking up the sentence, while the second does not. Try reading out these two versions. Parentheses can make your sentence clunky and hard to read out loud. Let's compare one more pair:

1. *Evolutionary theory predicts that animals will adjust their time spent foraging in different patches in response to the density of available food items (while controlling for predator density, as higher predator density will devalue patches), so as to maximize the rate at which they acquire nutrition (assuming that all food items provide equivalent energy returns).*

> 2. *Evolutionary theory predicts that animals will vary the time they spend foraging on a patch in response to both food availability and predator density. Individuals are favoured to maximize the rate at which they acquire food, and so will spend longer foraging in patches where food items are at higher density or higher quality. Individuals are favoured to spend less time foraging in patches where they are more likely to encounter predators.*
>
> In the second version, as well as removing parentheses, we divided the text into more and shorter sentences. This has increased the overall length, but it's worth it, because it's much clearer and easier to read. If you can, avoid parentheses.
>
> ### 3. Break rules—when it helps
>
> We know that we have said it many times, but we really cannot stress enough that you shouldn't see our advice as hard-and-fast rules that can't be broken. The bottom line is to do whatever you think will better help your reader.

Keep Working at Your Writing

Good scientific writing does not happen overnight. Even after reading this book! But, if you work at it, you will keep improving your scientific writing, throughout your entire career. Our aim has been to provide you with a toolkit that will act as a starting point (Cartoon 11.4).

Commenting on the papers of colleagues can be a useful way to learn about writing. By forcing yourself to be an active reader, who gives useful comments, you can see what has worked and what hasn't. You will see what makes for good reading, and what makes for frustrating reading. The same benefits can be obtained from reviewing papers.

There is a wealth of sources on writing, ranging from whole books[2] to web articles.[3] Sometimes these make different points, and sometimes they even disagree. But disagreement can be a good thing—remember that rules can be broken. What really matters is that different perspectives can help you think about how to be a better writer.

Writing up your research can be one of the most exciting and enjoyable parts of the scientific process. You have done your research, and now you get to tell everyone about it. Sure, writing can be a challenge, but it is a stimulating, thoughtful challenge that can

---

[2] There are a wealth of excellent books covering other aspects of writing, to help you keep thinking about and improving your writing. Strunk and White (*The Elements of Style* 4e, Pearson 1999) and Gowers (The Complete Plain Words, 1987) are both classics on writing style, focused on grammar, but a bit out of date. Our book is not a grammar guide – indeed one of us constantly needs their grammar corrected! Pinker (*The Sense of Style*, Penguin 2015) discusses scientific writing more broadly. There are several books on scientific writing that cover the more basic issue of how to write a paper, or examine the philosophy of paper writing in more detail, such as Cargill and O'Conner (*Writing Scientific Research Articles*, Wiley 2009), Heard (*The Scientist's Guide to Writing*, Princeton University Press 2016), and Schimel (*Writing Science*, OUP 2012). Hochburg (*An Editor's Guide to Writing and Publishing Science*, OUP 2019) explores the publishing process from an editorial perspective. Glasman-Deal (*Science Research Writing*, Imperial College Press 2010) is targeted at non-native English speakers. Orwell (*Why I Write*, Penguin 2004) discusses many aspects of writing, especially in his 1946 essay 'Politics and the English Language', where he introduces six rules for writing.

[3] There are a lot of free resources on the web. For example, search with 'scientific paper writing'.

Cartoon 11.4 *A toolkit for writing.*

be solved through planning, patience, and practice. The more you actively embrace that challenge, the more you will enjoy it.

Summary

Additional Advice for Polishing Papers
Active Writing
1. The different sections of the paper can be written in any order
2. Writing as you do the science can help
3. Plan where you will write, and when to take breaks

Edit, Edit, Edit!
1. Leave and forget
2. Be forensic
3. Remove unnecessary words/sentences/paragraphs
4. Maintain the narrative flow
5. Learn to love critics
6. Use whatever trick works

Keep Working at Your Writing